沙尘暴发生发展的观测和数值模拟研究论文集

姜学恭　主编

气象出版社
China Meteorological Press

内容简介

本书是作者沙尘暴研究论文的合集，涉及沙尘暴过程的大尺度环流特征和动力热力结构；影响沙尘暴的天气要素及其数值模拟研究；非绝热加热、地形等外强迫因素对沙尘暴发生发展的影响；地形影响沙尘传输的机理及概念模型；沙尘气溶胶的对流层—平流层输送特征等。

本书可以为从事天气预报、气候预测、大气科学研究以及自然地理、生态、资源与环境、灾害学、农牧业、林业、交通运输等领域的专业技术人员提供参考。

图书在版编目(CIP)数据

沙尘暴发生发展的观测和数值模拟研究论文集 / 姜学恭主编. --北京：气象出版社，2017.3

ISBN 978-7-5029-5191-7

Ⅰ.①沙… Ⅱ.①姜… Ⅲ.①沙尘暴-文集 Ⅳ.①P425.5-53

中国版本图书馆 CIP 数据核字(2017)第 025046 号

Shachenbao Fasheng Fazhan de Guance he Shuzhi Moni Yanjiu Lunwenji

沙尘暴发生发展的观测和数值模拟研究论文集

出版发行： 气象出版社

地　　址： 北京市海淀区中关村南大街 46 号　　**邮政编码：** 100081

电　　话： 010-68407112(总编室)　010-68408042(发行部)

网　　址： http://www.qxcbs.com　　**E-mail：** qxcbs@cma.gov.cn

责任编辑： 李太宇　　**终　　审：** 邵俊年

责任校对： 王丽梅　　**责任技编：** 赵相宁

封面设计： 博雅思企划

印　　刷： 北京建宏印刷有限公司

开　　本： 787 mm×1092 mm　1/16　　**印　　张：** 10.5

字　　数： 270 千字

版　　次： 2017 年 3 月第 1 版　　**印　　次：** 2017 年 3 月第 1 次印刷

定　　价： 50.00 元

前　　言

21世纪初，东亚沙尘暴出现了一个明显的多发期，期间，沙尘暴的尺度、发生次数、传输距离和影响范围等方面都接近历史之最。这也使得国内沙尘暴研究出现了前所未有的"繁荣"。相关的研究拓展到了沙尘暴的各个方面，包括沙尘暴天气和气候特征、地基和遥感监测技术、成因及形成机理、天气影响系统及气象要素特征、沙尘气溶胶理化特征及辐射强迫、沙尘暴环境、生态和气候效应、沙尘暴预报预测方法及数值预报技术。同时，人们也积极开展沙尘暴的综合治理工作。

之后，由于深入开展的荒漠化治理，以及更为重要的气候波动导致的冬季风强度减弱，东亚沙尘暴呈现波动减少的态势。在这种情况下，人们对于沙尘暴的关注度明显降低，沙尘暴相关的研究也明显减少。特别是近年来雾、霾天气的频发，极大地吸引了人们关注的目光。然而，沙尘暴作为发生历史更为久远的天气现象，目前的荒漠化治理与彻底根治沙尘暴灾害的最终目标之间还有漫长的路要走。而且，沙尘暴的危害绝不仅局限于对工农业生产、环境及健康的影响。沙尘气溶胶辐射强迫能够对大气热量平衡及气候变化产生深远影响，并通过与大气气溶胶之间的理化反应影响降水等更为复杂的大气过程，因而成为气候变化诸多影响因素中最为敏感的因素。这在气候变化对人类未来产生重要影响的今天，无疑是不应忽视的。

内蒙古西部沙漠和荒漠化地区是东亚沙尘暴的主要沙尘源地。该地区沙尘暴过程常由冷锋、蒙古气旋等天气系统引发，其过程强度强，传输距离大，影响范围广，往往对东亚地区产生严重影响。因而，沙尘暴过程历来都是内蒙古天气预报所重点关注的，也是包括作者在内的内蒙古天气预报工作者持之以恒开展分析和研究的课题。

本论文集是作者近年来在沙尘暴方面的一些粗浅认识，主要包括沙尘暴成因及天气要素特征，沙尘暴数值模拟研究及外强迫因素影响研究，沙尘气溶胶对流层—平流层输送等。希望能为认识和了解沙尘暴，以及预报和治理沙尘暴提供一些借鉴和参考。由于作者水平有限，错误和疏漏在所难免，希望广大读者批评指正。

作者

2017年2月10日

目 录

上编　沙尘暴影响因素及动力热力结构

Numerical Simulation of Synoptic Condition on a Severe Sand Dust Storm

JIANG Xuegegong[1] SHEN Jianguo[2]
LIU Jingtao[1] CHUNG Yongseung[3]

1. Inner Mongolia Meteorological Observatory, Huhhot, China;
2. Inner Mongolia Meteorological Bureau, Huhhot, China;
3. Korea-China Center for Atmospheric Reseach, National University of Education, Korea

Abstract By using observational data and MM5, the observational analysis and numerical study was conducted on the synoptic condition of a case of severe dust storm caused by Mongolia cyclone occurred from 6 April to 8 April 2001. The results shown that the cyclogenesis was owed to the isentropic potential vorticity(IPV) advection in the upper troposphere and the terrain modified baroclinicity in the mid-lower troposphere. The Altai-Sayan complex mountain blocked the lower level cold air and made the isentropic surface sharply steep. When the air slid down along the isentropic surface the increasing of baroclinicity and decreasing of stability blown up the vertical vorticity development.

The formation of the dust storm was owed to the cyclonic cold front going across this area. The occurrence of dust storm had close relation to the strong surface wind, which was accompanied by the cold front passing, rather than the cyclogenesis itself. The reason that the prefront dust storm formed was the formation of heating convection. The reasons of the formation of the black storm (visibility lower than 50 m), which occurred at the mid-north part of Inner Mongolia, lay in several aspects. Firstly, at this area the surface wind was strong, which was caused by the downward transport in mid-lower troposphere. Secondly, the cold front passed there near the sunset time and so the air obtained much more surface heating to form deeper mixed layer(ML). Thirdly, the cooperation of the lower level wind and the terrain made the atmosphere at this area acquired the maximum advective contribution to forming a deep postfront ML. The sensitivity experiment run shown that surface heat flux was important to the frontal lifting. The forcing of surface heating was also the primary forcing mechanism of frontogenesis. Meanwhile, removing of surface heat flux made the atmospheric stratification became stable and the prestorm ML was very shallow, which would waken the strength of the dust storm.

Key word: Mongolia cyclogenesis, dust storm, mixed layer, numerical simulation

1 Introduction

Form 6,April to 8,April 2001,a severe dust storm induced by the deepening of a Mongolia cyclone occurred in northern China. During the period of the development of Mongolia cyclone,the central pressure decreased from 1003 hPa to 977 hPa. A black storm(Fig. 1)occurred at the mid-north part of Inner Mongolian where 30783 animals were killed,30000 animals went astray and 3000 acres of grassland was buried.

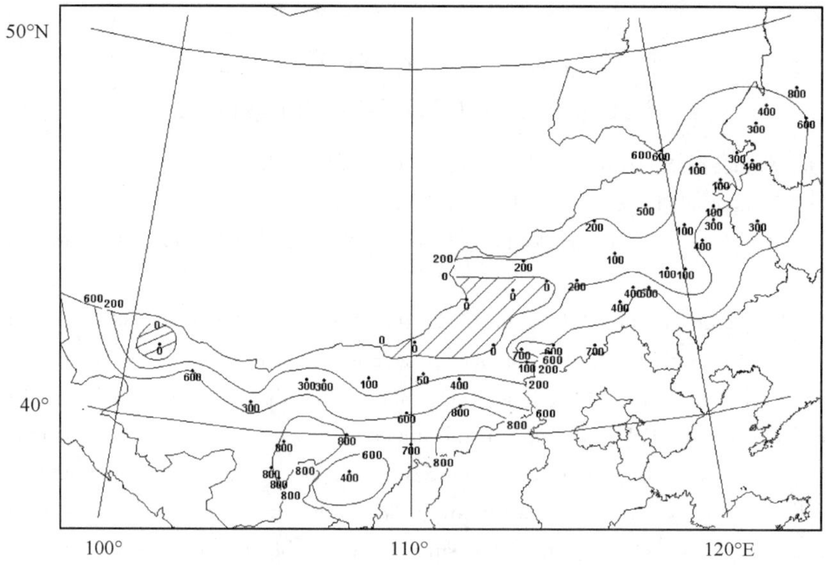

Fig. 1 The minimal visibility appeared in Inner Mongolia in this case(unit in m,interval 200 m and line of 400 m was omitted). The shadows were the black storm area(visibility equal to 0 m). The data came from the surface weather chart from 6 April,00 UTC to 7 April,12 UTC every 6 hours.

There were not adequate studies on the Mongolia cyclongenesis. The Altai-Sayan complex mountain plays an important role in most cases of the Mongolia cyclogenesis. These cyclones can be divided into two types,namely "terrain induced"(forcing at low-level atmosphere)and "terrain modified"[1,2] (forcing in upper-level atmosphere). Other factors such as the baroclinicity and stability are also important to the Mongolia cyclone development. On the reason for the formation of dust storm,some authors[3] emphasized the convective instability in determining the formation of dust storm besides the dry climatic background and the frontal activity. Hu[4] pointed out that the dust storm was always associated with a mesoscale low and dry squall line in northwest and north parts of China. Takemi[5] found out that the deep ML played a significant role in determining the formation of the dust storm that occurred in northwest of China on May 5,1993. In this present study a case of severe dust storm occurred from 6 April to 8 April 2001 is simulated by using Pan State/NCAR mesoscale model (MM5). The output data and some observational data are analyzed in order to find out the physics causing the development of Mongolia cyclone and blow up of the severe dust storm.

This process can be divided into two stages. From 5 April,02 UTC to 6 April,08 UTC, the cyclone approached Altai-Sayan complex mountain and its central pressure decreased from 1003 hPa to 991 hPa. In this stage the cyclone was comparatively weak. From 6 April, 08 UTC to 7 April,08 UTC the cyclone moved along the eastern slope of the mountain and rapidly deepened with its central pressure decreasing from 991 hPa to 977 hPa. In this stage, a large-scale dust storm occurred in northern China. From 6 April, 20 UTC to 7 April, 02 UTC when the cyclonic cold front went across Inner Mongolia, the dust storm extended to the middle-north part of Inner Mongolia and its intensity came to a peak. Here the black storm was observed at six stations (see Fig. 1). Afterward, the cyclone and the dust storm moved to the northeast part of China and their intensity started to weaken.

Figure 2 shown that the distribution of the maximal surface wind during this period was coincident with the area where the middle portion of cold front went across. The Maximal surface wind with up to 20 m/s was observed at three stations(such as HaiLiSu etc.). It was not coincident with the area of horizontal visibility wholly. The reason will be discussed in the latter sections. Figure 3 shown when the cold front and dust storm (at ErLianHot station) passed, the visibility evolution and the intensive change of the surface meteorological factors, including abruptly changing of wind, increasing of wind velocity, increasing of pressure, decreasing of temperature were similar to the dust storm induced by squall line[4,5]. But due to in the extreme dry environment, the relative humidity did not increase then.

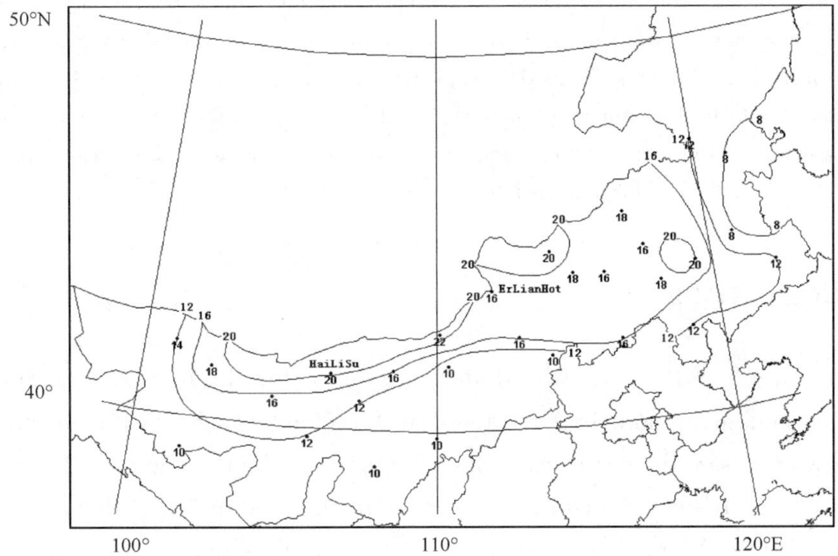

Fig. 2 The distribution of maximal surface wind speed in Inner Mongolia in this case. (unit in m/s, interval 4). The symbol ErLianHot(HaiLiSu) was the station where the black storm(dust storm) occurred. This data came from the surface observational data from 6 April, 00 UTC to 7 April, 12 UTC every 3 hours.

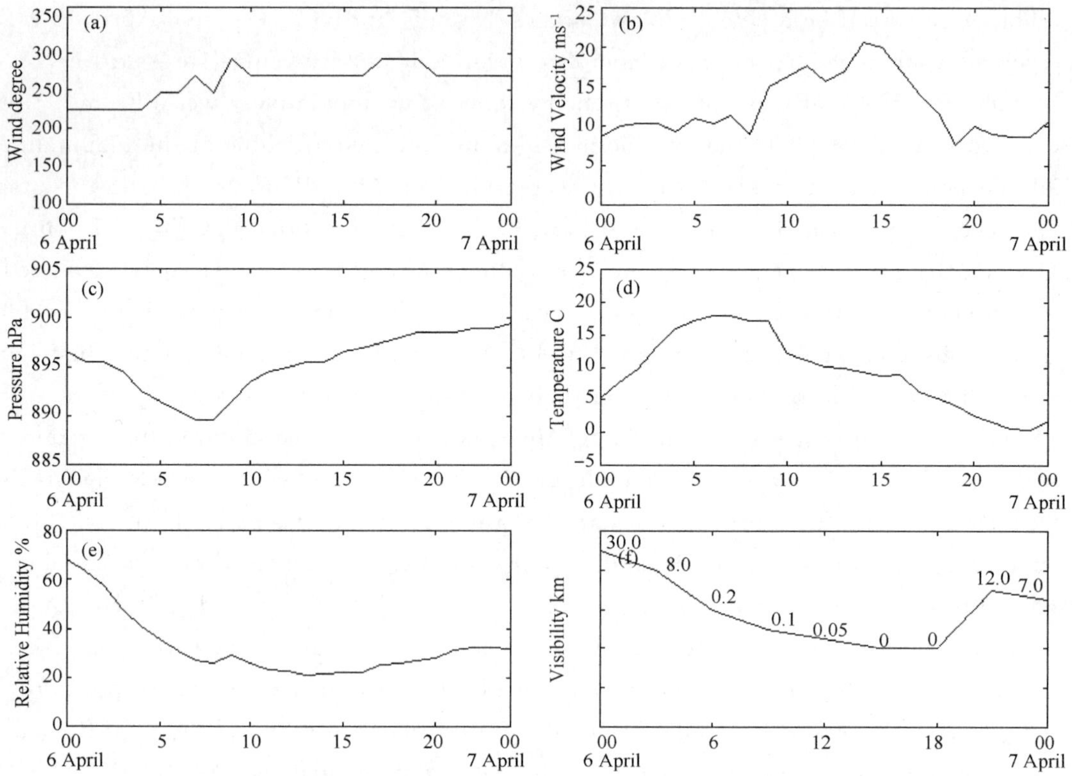

Fig. 3 The temporal evolution of surface meteorological factors, which included wind(a) unit in degree, wind speed(b) unit in m/s, pressure(c) unit in hPa, temperature(d) unit in ℃, relative humidity(e) unit in % and surface horizontal visibility(f) at ErLianHot station(6 April, 00 UTC to 7 April, 00 UTC, local time: 6 April, 08 LST to 7 April., 08 LST, interval 5 hours). The data came from the surface autographic record(The visibility data came from the surface observation every three hours. The dust storm was defined as visibility lower than 1000 m while black storm as visibility lower than 50 m).

2 Modeling setting

The Pan state/NCAR Mesoscale Model[6] was used for the numerical simulation in this study. The coarse domain(domain 1) centered at 45°N and 110°E. Its dimension was 41×45 and grid size was 90 km. Insides it, the dimension of nest domain(domain 2) was 73×73 and grid size was 30 km. Insides domain 2, the nest domain(domain 3 and domain 4) covered the cyclone area and Inner Mongolia(dimension is 139×76, grid size is 10 km). MRF PBL scheme with multi-layer soil model and cloud radiation scheme with radiation frequency of 30 minutes were used in control experiment(EXP CE). The PSU/NCAR 13 category Landuse data set and NCAR 30 min, 10 min, 5 min globe terrain data set were used for the four domains through 2-dimension interpolating. The HLAFS(limited area model; Beijing, China) analytic data were used for the first guess field at 11 pressure levels for 5 temporal periods(from 6 April, 00:00 UTC to 8 April, 0000 UTC, interval 12 hours). 50 hPa was defined as the P-

Top. The data were interpolated to 24σ levels as the initial condition. The simulation started at 6 April,00 UTC and terminated at 8 April,00 UTC.

Through removing the surface heat flux(sensible heat flux and latent heat flux)the sensitivity experiment(EXP NSFE)was conducted with all above other schemes contained.

3 Simulation Results

(1)The Development of Cyclone

The control run finely captured the temporal evolution of Mongolia cyclone and the cold front. From Fig. 4,we can see the modeling strength(and position)of the cyclone(and the cold front) is quite coincident with those of the observational data. They also indicated that simulations at the area(cyclonic center and cold front) covered by the nest domains were more accurate than around them. Additionally,the strength(29. 8 m/s versus 26. 0 m/s 12 UTC)and position of the modeling wind speed center at 850 hPa are similar to that of observational data.

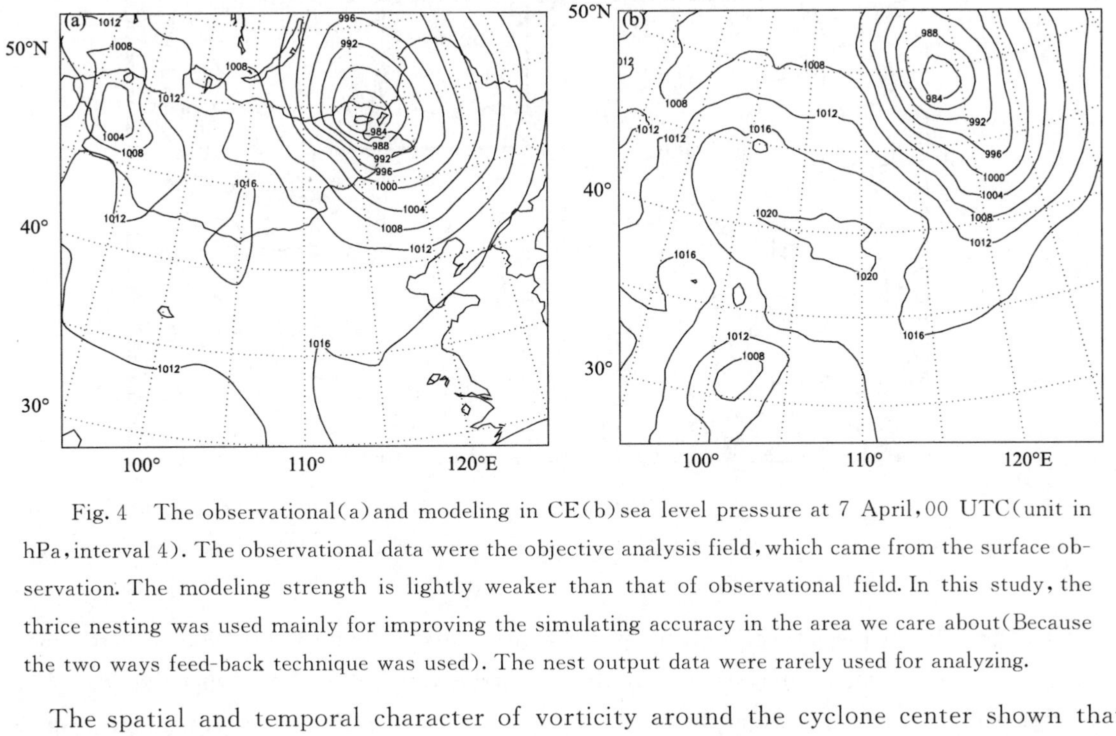

Fig. 4 The observational(a)and modeling in CE(b)sea level pressure at 7 April,00 UTC(unit in hPa,interval 4). The observational data were the objective analysis field,which came from the surface observation. The modeling strength is lightly weaker than that of observational field. In this study, the thrice nesting was used mainly for improving the simulating accuracy in the area we care about(Because the two ways feed-back technique was used). The nest output data were rarely used for analyzing.

The spatial and temporal character of vorticity around the cyclone center shown that there were two centers of large vorticity value appearing in upper and lower troposphere respectively. During the period of intensive cyclogenesis(6 April,00—12 UTC)the central vorticity at 750 hPa was larger than that at 300 hPa(16 versus 8 10^{-5} s^{-1}). The growing of upper-level vorticity was likely induced through the forcing in upper-level atmosphere where the IPV advection played a very important role. Figure5 shown that the high IPV(central value was 5. 26 PVU)expanded southwards to the west of cyclonic center(102°E). And the extended IPV was cut off there(112°E,52°N). Form Fig. 5 we can also see the IPV was strongly transported (advected) to the cyclone area by the west wind on the isentropic surface. Moreover, the cross section (N-S direction through IPV center, not shown) shown that the formation of the cut off IPV (112°E, 52°N) was caused by the downward and southward IPV transport from the polar and upper-level troposphere (The air there is always called IPV reservoir). Its mechanism was similar to the so-called tropopause fold. But the IPV was not transported to the cyclone center directly, so the tropopause fold was not significant[7,8]. The strongest transport occurred between 300 hPa and 400 hPa. So the IPV transport mainly impacted vorticity development in the upper level troposphere.

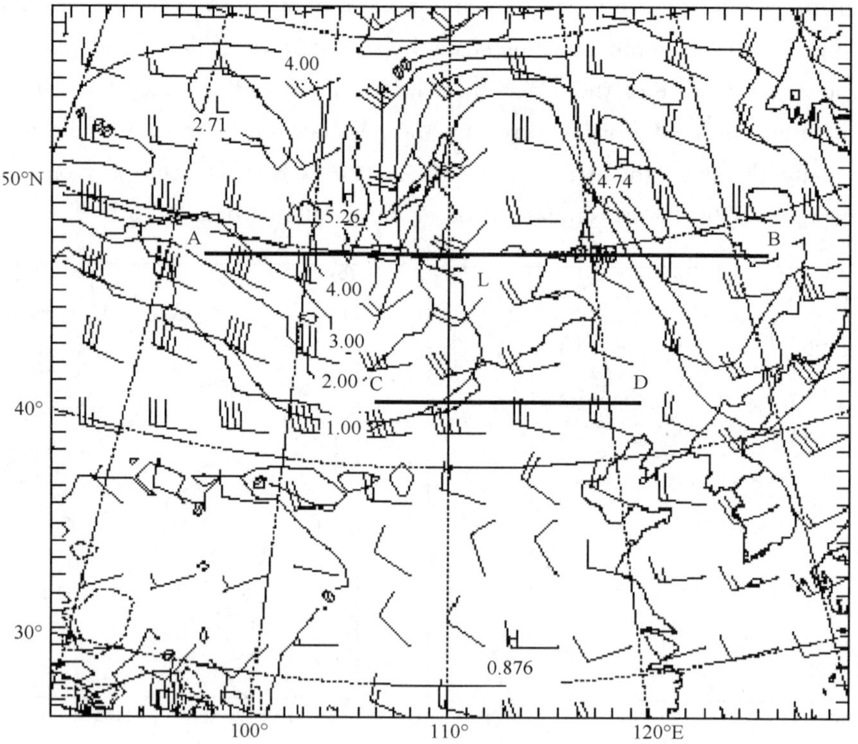

Fig. 5 The modeling (CE) potential vorticity (solid, unit in PVU, interval 1) and wind barb on 304 K isentropic surface (6 April, 09 UTC). The symbol L represents the cyclone center. The line AB represents the cross section position in Fig. 6 (domain1). The line CD represents the cross section position in Fig. 9, 11 (domain 3).

The growing of lower-level vorticity was mainly caused through the forcing of baroclinicity, which was obviously modified (enhanced) by the topographty. Figure 6 shown that the 288—304 K isentropic surfaces slanted down at the nearby west of the cyclone that was the east edge of the Altai-Sayan complex mountain. Moreover, the wind field on the isentropic surface shown that the air at this area would slide down along the slantwise isentropic surface due to the wind (not shown). According to the slantwise vorticity development (SVD) theory[9]

when a parcel slides down along the slantwise isentropic surface, the change of the baroclinicity and stability would cause the intensive development of vertical vorticity.

In the complete vertical vorticity equation $F_{cve} = \rho\partial/\partial t(P/\theta_z - C_D)$, which was put forward by G. X. Wu (2001, here P is the potential vorticity), some dynamical and thermal factors were combined into the parameter C_D to consider their effects on the vorticity development. When a parcel slides down a slantwise isentropic surface, if $C_D<0(C_D=\xi_s\theta_s/\theta_n)$ and C_D decreased, the vertical vorticity would develop intensively. (Here, ξ_s, θ_s and θ_n represents vertical wind shear, horizontal baroclinicity and vertical static stability, respectively).

Figure 6 shown the slantwise isentropic surface at the mountain's east edge was extreme steep and much more obvious in the lower troposphere. The strong North-South vertical wind shear also existed there. So $\xi_s>0$, $\theta_s>0$ and $\theta_n<0$ (which could be gained by decomposing the unit vector of θ to the horizontal component θ_s and vertical component θ_n along the outside normal direction of the slantwise isentropic surface). Thus, $C_D<0$. When the parcel slide down, the increasing of horizontal baroclinicity (θ_s) and the decreasing of vertical static stability (θ_n) determined the decreasing of C_D and so caused the intensive development of the vertical vorticity. It can be see that the cyclogenesis was accomplished by combining the effect of vertical wind shear, baroclinicity and static stability.

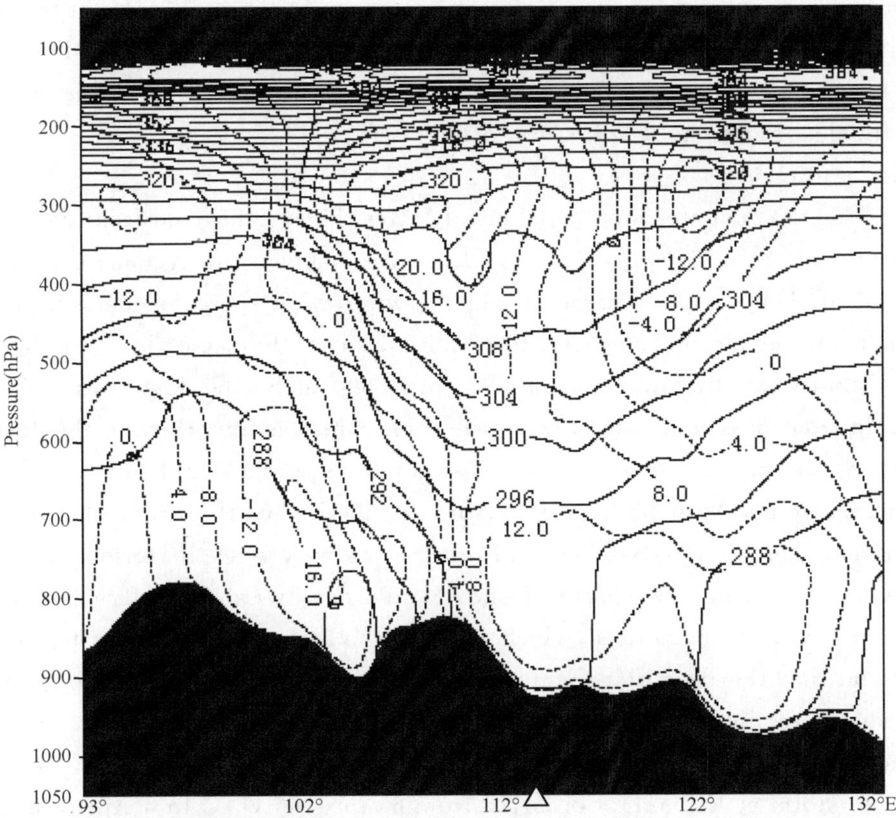

Fig. 6 The W-E direction cross section(in domain1) through the cyclone center at 6 April, 12 UTC (the line AB in Fig. 5). θ(solid, unit in K, interval 4) and N-S wind (dashed, unit m/s, interval 4). The positive (negative) values were south (north) wind. Around the triangle was the cyclonic area where the cyclone had the warm core structure. Below 300 hPa the isentropic surface slanted down on the west of cyclone (112°E, , east edge of A-S mountain) where the N-S wind strong vertical shear existed.

In this process the topographhty of the Altai-Sayan complex mountain enhanced the baroclinicity in the lower troposphere. From Fig. 6 we can see that due to the mountain retarding effect, the lower-level cold air was blocked at the east part of the mountain and could not quickly go across the mountain area[1]. And so it made the horizontal gradient of θ stronger and slantwise isentropic surface steeper than without it. According to the SVD theory discussed above, it(the enhancement of baroclinicity) would induce more extremely growing of vertical vorticity. (A sensitive experiment that lowering the elevation of Altai-Sayan complex mountains to 1/2 would induce the weakening of the cyclone obviously, not shown). This mechanism may be a manner with which the effect of lee performs on the lee cyclongenesis.

Comparing the vertical distributing charactor of θ_e with that of θ, we can see below 600 hPa the θ_e surface was certainly steeper than θ surface. It indicated that the humid baroclinicity and convective stability also played roles in cyclogenesis at a certain extent.

(2) The Formation of the Dust and black storms

In this case the minimal visibility(black storm) situated at the strong surface wind area, but did not just located at the maximal surface wind area(see Fig. 1, 2). So in this section the characters of black storm occurred at Erlianhot and dust storm occurred at HaiLiSu will be discussed and were used to illuminate the different character of the synoptic condition connected to the dust storm and black storm in this case.

1) Dust storm

In this case the dust storm occurred mainly after the cyclonic cold front passed(from the weather charts and cloud images, not shown). Figure 7 shown the cyclonic cold front passed at 6 April, 03:00 UTC(approximately) when the wind, wind speed, temperature and relative humidity abruptly changed. The dust storm initialized at the same time(nearby 03 UTC) when the visibility rapidly lowered from 8000 m to 300 m. So the dust storm initialized just after the cold front passed and it reflected the cold front was important to the dust storm.

In fact, it was the strong surface wind that did directly relate to the dust storm. The dust storm occurred from 6 April, 03 UTC to 6 April, 09 UTC when the strong surface wind speed appeared(larger than 15 m/s, See Fig. 7). From Fig. 8a we can see its formation was owed to the west wind momentum transported from 700 hPa downward to surface. There also were two characters worth to be noticed, which one was the ascending motion in this process was very weak and the other was that the dust storm began to weaken when the descending motion began to appear in the whole troposphere.

2) Black Storm

The dust storm at ErLianHot occurred from 6 April, 06 UTC to 6 April, 18 UTC while the black storm from 6 April, 12 UTC to 6 April, 18 UTC. Meanwhile the cold front passed at 6 April, 09 UTC(See Fig. 3). So we can see a major character at ErLinHot different with at HaiLiSu was the dust storm also occurred when the cold front had not come(from 6 April, 06 UTC to 6 April, 09 UTC). From the weather charts(not shown) we can see the prefront dust storm appeared only at the black storm area(mid-north part of Inner Mongolia) and appeared only near 6, April 06 UTC, which implied that it was not a universal pattern. Figure 3 shown that within this period the surface wind was S-W wind and the wind speed was very steady (not strong, about 10 m/s). Meanwhile there was increasing of temperature, decreasing of surface pressure and increasing of T-Td obviously. Figure 8b also shown that the ascending motion below 600 hPa was much stronger than at Fig. 8a (HaiLiSu) within this period. So there likely was a warm low forming and the dust storm seem to be induced by the local heating convection and the dust came from the local raising (This regions was the retrogressive grass land).

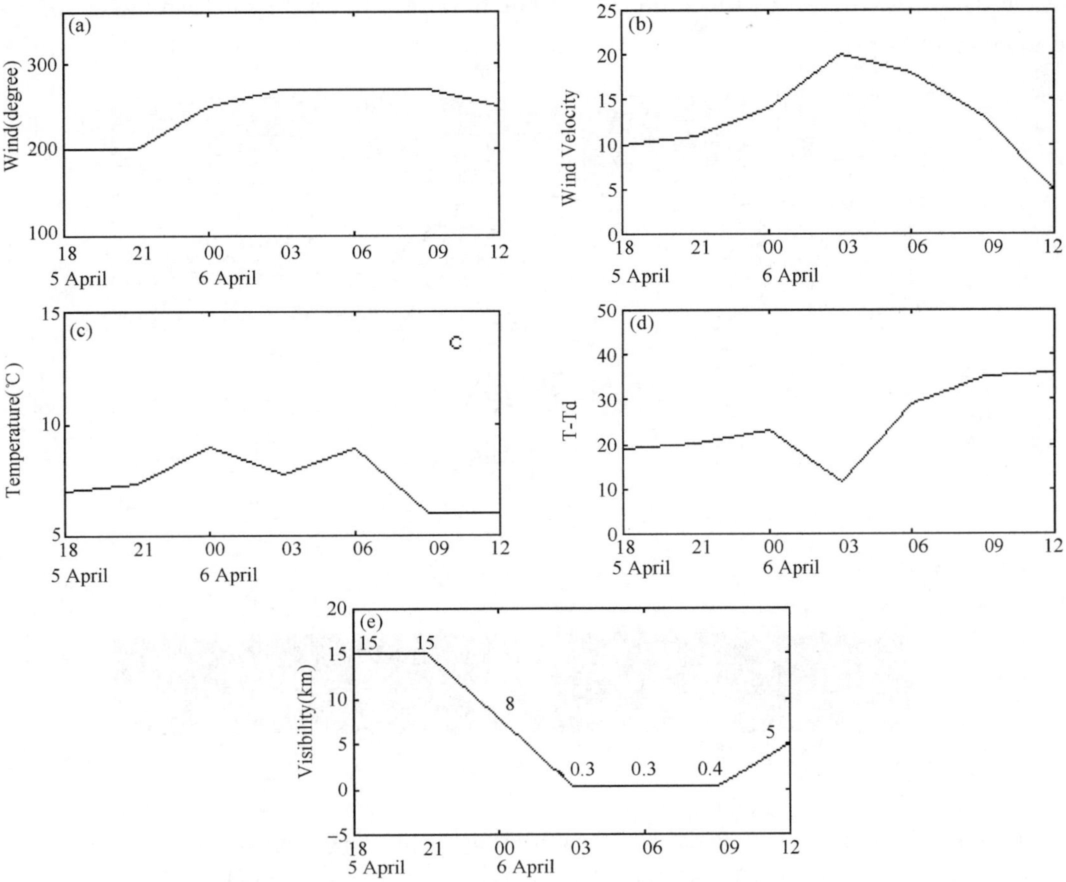

Fig. 7 The temporal evolution of surface meteorological factors like Fig. 3 except (1) at HaiLiSu (see Fig. 2) (2) without surface pressure (3) difference between temperature and dewpoint temperature ($t-t_d$) rather than relative humidity (RH) (4) period from 5 April, 18 UTC to 6 April 12 UTC (5) The data came from the surface observation every three hours.

Like at HaiLiSu, the *severe* dust storm had also close relations with the cold front and its induced strong surface wind. Figure 3 shown that the *dust storm* started to enhanced just after the cold front passed (visibility equal to 100 m, 09 UTC) and the *black storm* initialized just after the first peak of surface wind speed appeared (visibility equal to 50 m, wind speed equal to 18 m/s, 12 UTC) and came to a head just after the second peak of surface wind speed appeared (visibility equal to 0 m, wind speed equal to 20 m/s, 15 UTC). So, like at HaiLiSu, the black storm initialized after cold front passed and the severe black storm(visibility equal to 0 m)later than the appearing of the peak of surface wind speed. Here, there was a remarkable character difference with at LaiLiSu, which was that the ascending motion was greatly strong both of before and after the cold front passed. (Figure 8b shown that after the cold front passed the ascending motion induced by the frontal lifting extended up to 300 hPa accompanying by the frontal moving.)It reflected the cold front was extremely stronger than at HaiLiSu and its effect on black storm will still be discussed latter. Even though it, at ErLianHot and HaiLiSu, there was the similar relations between the cold front, strong surface wind and dust(black)storm.

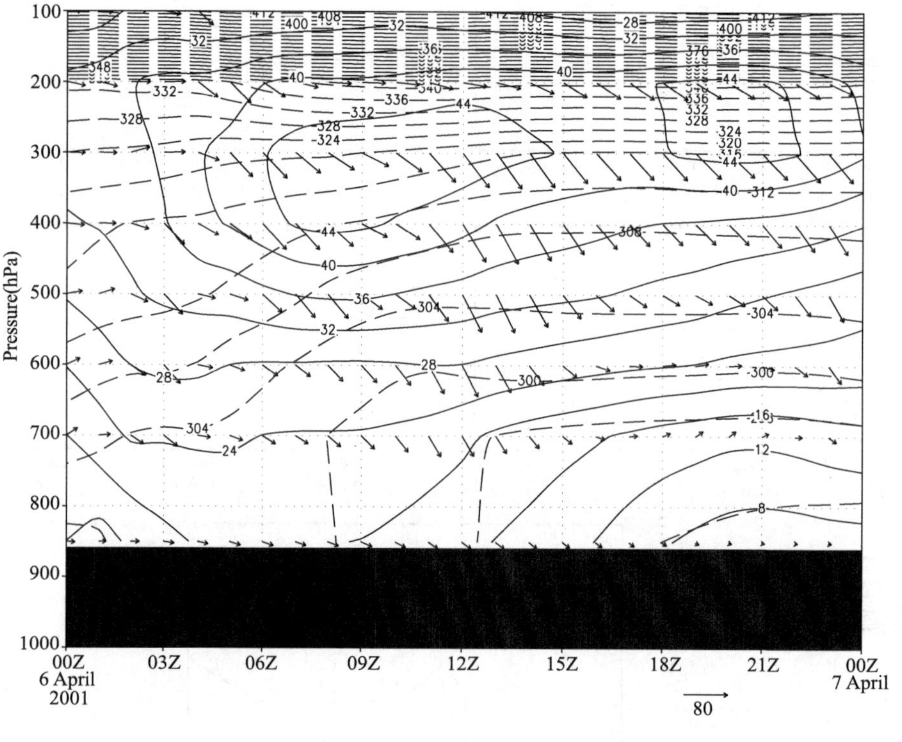

Fig. 8a The temporal evolution of W-E wind(solid, unit in m/s, interval 4), θ_e (dashed, unit in K, interval 4)and U+W wind vector(U, m/s and W, 0.001 m/s)at HaiLiSu station in CE.

3)The formation of ML

Besides the strong surface wind and sand source factors, the dry deep convection was crucial to the formation of severe dust storm[3,5]. The deep dry convection can provide fully mixing of air and so can raise and suspend the sand-dust up to middle troposphere. Moreover, it can transport the upper-level momentum downward to surface. The formation of deep convection was generally caused by the convective instable stratification of atmosphere. However, at the arid and semiarid regions the instable stratification could not be expected due to the small amount of moisture[10]. In this case the air on the south of 45°N was very dry and no precipitation was produced(can also see Fig. 3 and Fig. 7). Although the instable stratification was not satisfied in this case, the formation of deep mixed layer and its special stratified structure in the prestorm was favorable to the formation of deep dry convection.

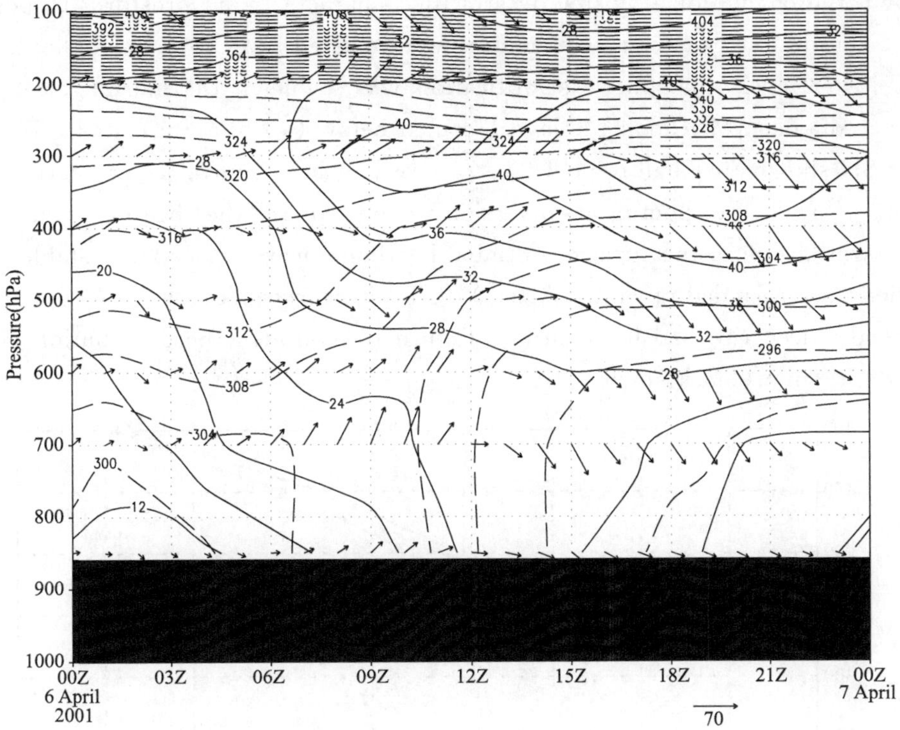

Fig. 8b The temporal evolution like Fig. 8a except at ErLianHot.

Gamo[10] suggested that there often is deep ML developing up to 4～6 km in the daytime above some deserts in Asia and Africa. The effect of deep ML on the formation of dust storm likely performed at two aspects. Firstly, in an extremely dry environment the level of free convection(LFC)is very high, so lifting air above the LFC requires much more energy(convection required). The deep ML can decrease the difference in height between the top of ML and the LFC(minimizing the lifting energy)and so favorable for the formation of deep convection[5]. Secondly, (1)the higher the atmospheric layer in convective condition extended to, the higher could the sand-dust be raised and suspended(beside the factor of particle size). (2)When the atmospheric stratification was in the convective condition, the dust concentration approaches a constant value with height[11]. (3)The deep ML can act as a character(or a result)of convective condition. So in the deep ML condition the dust could be suspended very high and transported to a long-distance and so the storm was severe. So, we can regard the height of ML as a character of dust storm, which can indicate its strength.

In this case, The ML had a special stratified structure like Gamo's study[10]. Gamo found that at arid or semiarid regions there was a weakly stratified, near-neutral layer at the mid-upper tropospheric layer(4～6 km), which did not change daily. Below it the atmospheric stratification was stable at night due to the cooling of surface radiation and became near-neutral in the daytime due to the surface heating. Gamo thought that the deep daytime ML developed from the surface ML linking with the existing upper-level near-neutral layer. Like that, Fig. 9 shown the change of atmospheric stratified character from prestorm to poststorm at ErLianHot station. We can see a near-neutral layer existed from 400 hPa to 300 hPa and below it the lower level atmospheric stratification was stable in the morning. But after dust storm occurred a deep ML(adiabatic layer)formed below 600 hPa(estimated). The modeling result(Fig. 8b) shown that before 03 UTC there was a near-neutral layer nearly at 400 hPa and below it was the nocturnal inversion layer where the heating convection was weak. Afterwards, due to the surface continual heating a near-surface weak stable layer gradually formed and expanded upwards. After 06 UTC it nearly linked with the upper near-neutral layer and a deep ML had been formed. Figure 10 exhibited the distribution of prestorm ML around Erlianhot(black storm).

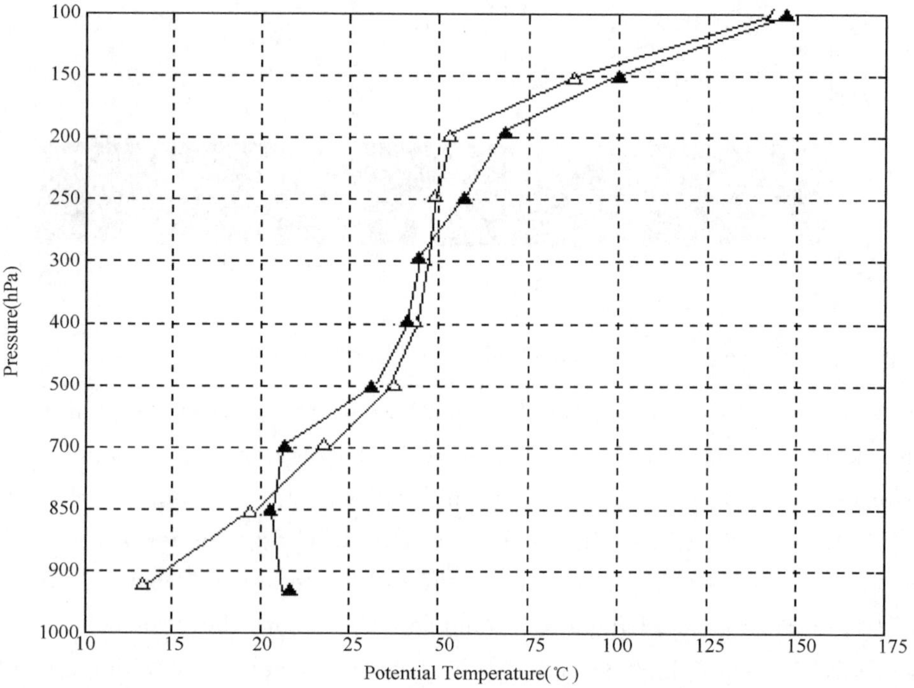

Fig. 9 The prestorm(00 UTC, in the morning, hollow triangle) and poststorm(12 UTC, near the sunset time, black triangle) sounding at ErLianHot station. The x-coordinate represents potential temerature(uint in ℃, interval as 5 when θ smaller than 25 or 25 when θ larger than 25). The y-coordinate represents pressure(unit in hPa).

Although the two near-neutral layers can not really link together(Because the surface heating in the daytime could not reach the height of 2～3 km above surface due to the small latent heat release in a dry environment[10]), the upper near-neutral layer was easily mixed by some disturbance. It has been ascertained by the height of the heavy dust and oil smoke layer that often extended to about 500 hPa[10,12]. When the cold front passed at 09 UTC the frontal lifting(Because the small lifting energy required)overset the restrain of stable layer(between the two near-neutral layer)and intensive deep convection was triggered. Thus, the severe dust storm was induced. Comparing Fig. 8a and Fig. 8b we can see the great difference between at

ErLianHot and at HaiLiSu. The deep prefront ML formed up to 500 hPa at ErLianHot, which was owed to the strong surface heating, while only up to 700 hPa at HaiLiSu. Moreover, after cold front passed the ascending motion at ErLianHot was much strong than at HaiLiSu. Therefore, we can see the deep ML induced by the surface heating was an important reason that the black storm occurred at around ErLianHot and it acted as its first operating manner. Furthermore, because after front passed the great difference of ascending motion between them indicated the obvious difference of the cold frontal strength and structure. So the surface heating, which induced the postfront heating convection and deep ML and so made the cold front acting as a deep and strong system, may be the main forcing mechanism of cold frontogenesis.

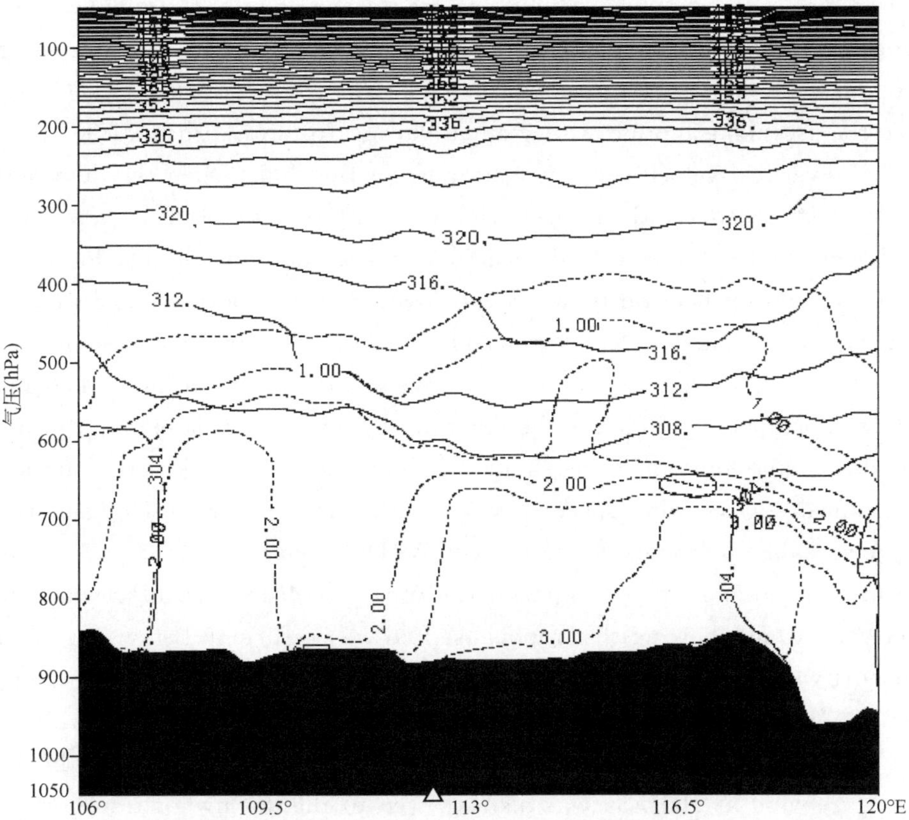

Fig. 10 The W-E cross section(in domain3) through black storm area(the line CD in Fig. 5, 07 UTC, one and half hour before front passed at ErlianHot) in CE. θ_e (solid line, unit in k, interval 4) and Q_v (mixing ratio, dashed line, unit g/kg, interval 1). The coupling layer of strong vertical gradient of θ_e and Q_v represents the top of ML. Here we used θ_e substituted θ because the difference between them was small. The triangle represents the ErLianHot station.

The daily change of ML also impacted the appearing area of the black storm. Although from 100°E to 110°E(See Fig. 2) the surface wind was also very strong and some deserts(BaDanJiLin, Wulanbuhe and Kubuqi Desert) existed there, which were favorable for the formation of severe dust storm, the black storm appeared only on east of 110°E rather than just at there(like HaiLiSu). Comparing Fig. 8a and Fig. 8b we can see the reason was that on the east of 110°E the cold frontal passed(08:30 UTC—11:30 UTC,16:30 LST—19:30 LST Beijing local time)at near sunset(latter than the desert regions for about 6 hour)when the lower atmosphere had been sufficiently heated and the surface ML had been well developed(the ML reached peak at sunset)and the frontogenesis had come to a head. But above these deserts it did not do due to the short time of surface heating(like HaiLiSu). So, despite the surface wind was also very strong,there was only one station's(EJiNaQi)visibility lower than 50 m (left center in Fig. 1). So the daily change of ML may be the most important reason that the black storm occurred just in the mid-north part of Inner Mongolia.

4)Downward transport of momentum

Based on the conclusion above we can analyze the character of the downward transport of momentum and the formation of strong surface wind. From Fig. 8b we can see the strong downward transport of west wind momentum appeared from 6 April,09 UTC to 6 April,12 UTC when the ascending motion was dominative,and from 6 April,12 UTC to 6 April,15 UTC when the vertical velocity was not clear below 500 hPa. When the descending motion dominated in the troposphere the surface wind started to weaken. In Fig. 8a it had the similar character. It implied the large-scale descending motion was not suitable for the downward transport of momentum. Contrarily,when the downward transport of momentum occurred a deep ML formed in the middle-lower troposphere in Fig. 8b. Fig. 8a was similar to it except the ML was lower correspondingly. So we can see the formation of deep ML caused by the heating convection was favorable for the downward transport of momentum rather than the large-scale descending motion. It was similar to some results given by other authors[13].

The maximal surface wind speed was 20 m/s at both of ErLianHot and HaiLiSu. But there was remarkable difference between them. At HaiLiSu(Fig. 8a)the downward transport mainly occurred in the mid-upper troposphere(from 700 hPa to 400 hPa)where the vertical wind shear was strong showing the tropopause fold was sufficient. Below 700 hPa the downward transport was very weak. On the contrary,at ErLianHot(Fig. 8b)in mid-upper troposphere the tropopause fold and the vertical wind shear were very weak while the strong downward transport occurred below 600 hPa where the deep ML formed. So the formation of maximal surface wind at HaiLiSu was mainly owed to the downward transporting in upper troposphere(tropopause fold)while at ErLianHot owed to the downward transporting in lower troposphere where the Deep ML was deep and favorable for it.

Although the strong descending motions did not closely related to the strong surface wind,it likely acted as a character of postfront dust storm in this case. Like at HaiLiSu,the severe black storm initialized just after the strong descending motions appeared. (Figure 8b shown the ascending motion below 500 hPa was completely replaced by the strong descending motion from 15 UTC to 18 UTC when the severe black storm occurred). In the transporting (then forming dustfall)dust storm process,the descending motion was always the dominative function[14]. From the cloud images(not shown)we can got a glancing idea that the sand-dust layer mainly came from the deserts in southern Mongolia. Thus, we can infer that the post-front black storm in Inner Mongolia likely came from the transport. But, when the descending motion continually strengthened, the storm started to weaken(the close relation maintained a short time). On the other hand, the prefront dust storm came from the local raising, and the ascending motion was also important to the strength of postfront storm. So the local raising likely contributed to the dust process. From the closely accompanying between the postfront deep ML and postfront dust storm(see Fig. 13), we can infer that the sand raising and dust-fall likely happened simultaneously at this area. It was still worth to study profoundly.

(3) The NSFE Result

1) The effect on cyclogenesis

From discussed above we can see the thermal forcing in the PBL played an important role in the formation of severe dust storm. So removing the surface flux(sensible heat and latent heat) we conducted the EXP NSFE. The results shown the weak effect appeared on the cyclogenesis(981 hPa versus 983 hPa, 7 April, 00 UTC,). It may be owed to the primary areas of surface heating were on the south of 45°N and mostly centralized from 100°E to 110°E rather than the cyclone area(desert ground character affected it). From the vertical distribution of vorticity at cyclone area we can see only at lower layer(maximum at 750 hPa) the vorticity was weaker than that in EXP CE. So, the effect of surface heating was not sufficient on the cyclogenesis within a short time and it was similar to those in some cases investigated by other authors, in which the surface flux had no efficient influence on the cyclogenesis for 24 hours simulation[15].

2) The effect on frontogenesis and dust storm

The surface flux had fatal effect on the cold front. The cold frontal vorticity distribution shown that the positive vertical vorticity only appeared below 500 hPa within the front passing period in NSFE. Figure 11 shown that after 09 UTC(front passed) no ascending motion appeared below 500 hPa. So, we can see removing the surface flux made the frontal lifting not available. It also validate the prefront heating forcing was the primary forcing mechanism of frontogensis.

The change of frontal lifting function was produced by the breakage of the balance between the cold and warm air and the variety of the atmospheric stratification. In Fig. 11 we could see before the cyclonic cold front passed the atmospheric stratification became very stable. Because removing the surface heating made the heating convection absent. And the formation of stable stratification restrained all types of convection growing(such as the frontal lifting, the mechanical forcing turbulent etc.). So the change of stratified stability made the frontal lifting not available after cold front passed and made the heating convection and deep ML absent before the clod front passed. Figure 12 shown that in NSFE the near-neutral layer (ML) did not appear above the black storm regions(from 110°E to 115°E). Although we could not see the change of dust storm in NSFE, through the change of ML and cold frontal lifting we could know that the strength of dust storm must had been greatly weakened.

There were two aspects worth to be noticed in Fig. 11. One is the ascending motion still appeared above 400 hPa after front passed. It indicated the effect of surface heating on atmospheric stratification appeared only below 400 hPa. Another was before 06 UTC the ascending motion from 700 hPa to 600 hPa in NSFE was slightly stronger than that in CE. The possible reason might be the N-S wind velocity and its vertical shear near 700 hPa became stronger than that in CE(not shown). And so the enhancement of wind shear around the top of PBL would form the mechanical forcing turbulent, which was favorable for the formation of forcing convection[16]. The reason for the enhancement of wind shear may be that removing the surface heating would decrease the heating convection and then weaken the vertical mixing in the PBL. So, we can infer that the effect of removing surface flux likely included two reverse processes. On the one hand, it eliminated the heating convection and the heating mixture. On the other, it enhanced the vertical shear that may strengthen the turbulent mixture induced by the mechanical forcing. Since the heating mixture was much greater than the mechanical mixture(exponential versus linear with time[17]), it finally induced the formation of strong stable stratification. It is no doubt that the heating mixture was crucial to the formation of deep ML and severe dust storm.

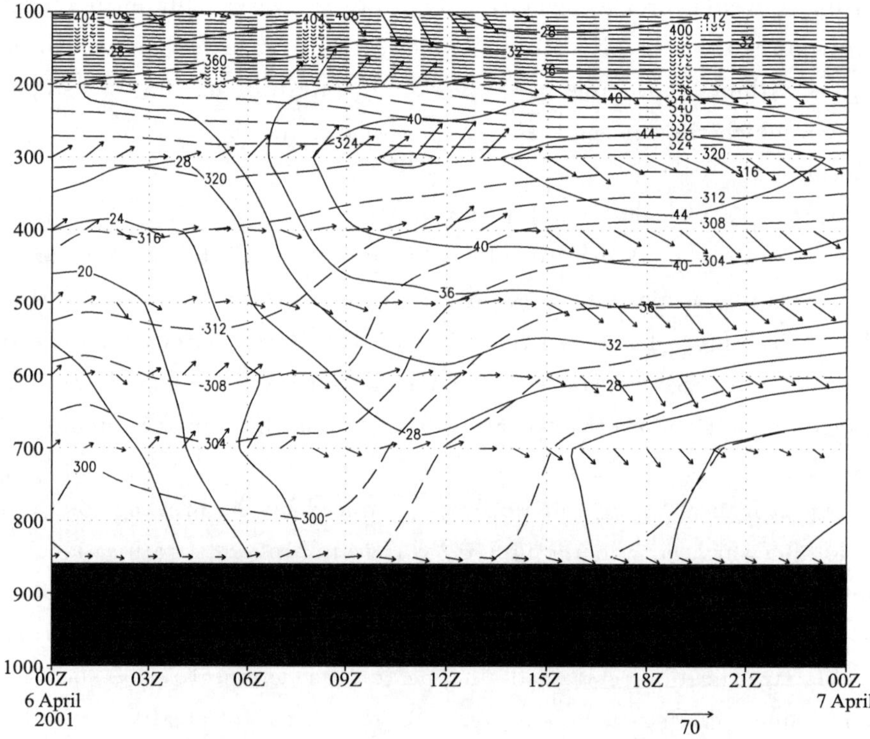

Fig. 11　The temporal evolution like Fig. 8b except in NSFE. Before 09 UTC the stratification was very stable. After 09 UTC, No available ascending motion appeared.

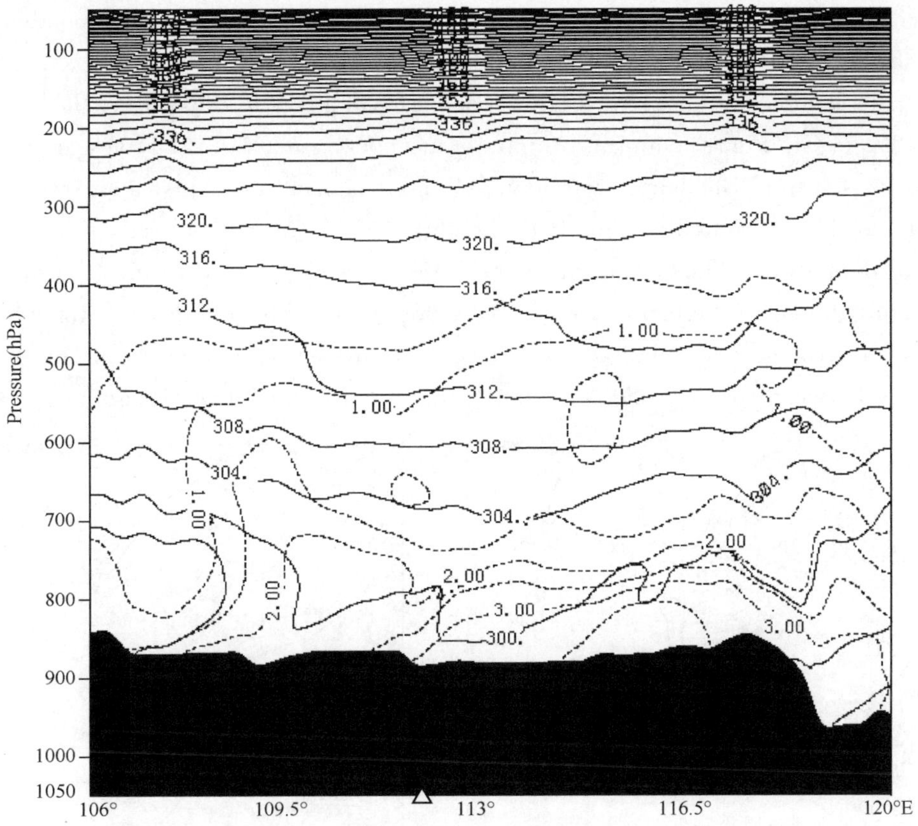

Fig. 12 The cross section like Fig. 10 except in NSFE. There was not the mixed layer appearing.

3) Another factor affecting on the severe storm

Authors had a finding in the CE worth to be noticed. Firstly, because the ML height was in accord with the height of PBL in the daytime[6], we could use the height of PBL to substitute the ML height. Secondly, from the distribution of PBL and terrain elevation in this case we can see they were similar partly because the terrain elevation can impact the height of PBL. Moreover, the PBL height was also partly impacted by the surface heating and it was easily comprehended. But we also found that in CE there was a series of high PBL located at from Sayan→west Mongolia→the mid-north part of Inner Mongolia(not shown). This distribution was not wholly in accord with the distribution of elevation, nor wholly in accord with the distribution of surface heating. This pattern was much more obvious in NSFE. Comparing it with wind field at 850 hPa we can see this series was greatly in accorded with the axis of lower-level jet. Because the lower tropospheric wind can advect the air of ML and it can impact the stratification of downstream air[12,16,18]. So we knew this series was induced by the advection through the lower tropospheric wind from the Altai-Sayan mountain area with the high elevation and high PBL to the middle part of Inner Mongolia. This effect was connected to the relative elevation of topographty between the two places(The effect of advection on the leeside atmospheric stratification determined by the different height between the mountain and the lee[19]). That is to say the middle part of Inner Mongolia just locates at the downstream area of lower tropospheric wind(N-W wind)and the difference between the heights of them was fit, so the effect of advection could be achieved. In order to prove it, we conducted an experiment that cut the elevation of Altai-Sayan complex mountains to constant(=1000 m). Comparing the PBL in CE(Fig. 13a)and in no mountain experiment(Fig. 13b), we can see after the cold front passed the PBL height was lowered about 800 m and it just appeared at the mid-north part of Inner Mongolia. The PBL height after front passed can partly reflect the strength of dust storm(The second manner of ML effect, discussed formerly), so the advection also contributed to the formation of severe dust storm. Meanwhile, the effect was accomplished by the proper mean wind at lower level, the proper relative position and elevation of topographty between the Altai-Sayan complex mountains and Inner Mongolia.

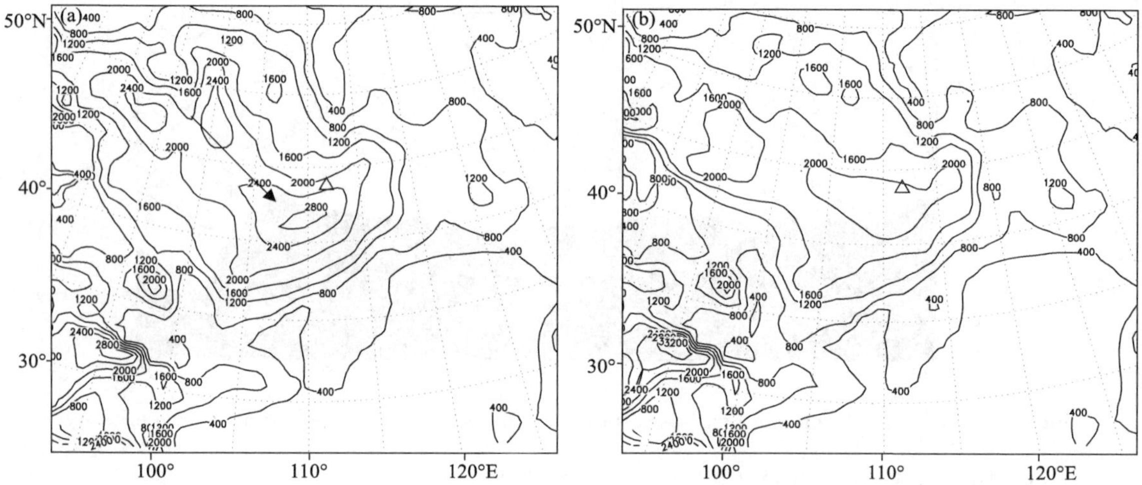

Fig. 13 The PBL height(unit in m, interval 400) at 12 UTC in CE(a) and no mountain EXP(b). The arrow vector represents the lower level jet(850 hPa). The black triangle represents ErLianHot station.

4 Conclusions

By using observational data and MM5, the observational analysis and numerical study was conducted on the synoptic condition of a case of severe dust storm caused by Mongolia cyclone occurred from 6 April to 8 April 2001. The main results included as follows:

(1)MM5 successfully simulated the development of Mongolia cyclone, the structure and the temporal evolution of the cyclonic cold front. Both of the isentropic potential vorticity (IPV)advection in the upper troposphere and the terrain modified baroclinicity forcing in the mid-lower troposphere were import in cyclongenesis. The baroclinicity forcing mechanism likely acts, when air slid down along the slantwise isentropic surface the increasing of baroclinicity and the decreasing of stability will blow up its vertical vorticity. Moreover, the Altai-Sayan complex mountains obviously blocked the lower tropspheric cold air, enhanced the baroclinicity forcing and so contributed to the cyclongenesis.

(2)This dust storm(like at HaiLiSu)in this case was caused by the cyclonic front going across this area. It closely related to the cold front and its induced strong surface wind. The formation of strong surface wind was caused by the downward transport that was important in the mid-upper troposphere and can be regarded as the tropopause fold.

(3) At the mid-north part of Inner Mongolia(like at ErLianHot), due to the strong surface heating a deep ML developed, which was accompanied by the heating convection that induced the prefront dust storm. The deep ML also contributed to the formation of postfront black storm because it minimized the required lifting energy and so favorable for the formation of dry deep convection when the cold front passed. The reason that the deepest prestorm ML formed in mid-north part of Inner Mongolia was the cyclonic cold front passed there near the sunset time. And so this area obtained the largest surface heating. It was just the daily change of ML induced the black storm occurred there rather than other areas. Meanwhile, the special structure of ML, which had an existing upper near-neutral layer not changing daily and a daily changed surface heating ML, was also favorable for the formation of deep convection and black storm.

(4) The formation of the black storm was also owed to the strong surface wind appeared after the front passed, which was caused by downward transport in the mid-lower troposphere. The transport was caused by the fully mixing in the deep ML rather than the tropopause fold. Moreover, another reason that the black storm occurred in mid-north part of Inner Mongolia was the strong advection of (isentropic) ML air from the Atai-Sayan complex mountains concentrated on this area and so it made the ML height(mainly at postfront) was much higher there than other areas. (Here, authors think the postfront ML height, which can represent the mixed grade of particles, may be regarded as a factor that can indicate the strength of the storm at a certain extend). And the advection effect could be accomplished only by combining the proper lower level wind, proper relate position and elevation of topographty between the upstream and downstream areas.

(5) The surface flux(sensible and latent heat) had weak effect on the cyclogenesis. But it was import to the strength of the cold frontal lifting. The strong frontongenesis at Inner Mongolia was mainly owed to the forcing mechanism of surface heating. It also was crucial to the formation of prestorm ML and the strength of the dust storm because removing the surface flux made the heating convection absent.

(6) The sand-dust storm was obviously a planet boundary process evidently. When the simulating and predicting are to be performed we should focus on the accurate simulation of the physical process in the PBL. All of the following factors, such as the PBL physical scheme, the surface heat flux, the accurate represent of terrain elevation and the precise wind represent in the PBL are very import to the formation of dust storm. Meanwhile the surface heating forcing was an important mechanism of frontongensis that worth to be carefully noticed.

Acknowledgements

The authors are grateful to Professor Shou-Jun Chen of Peking University for his directions and helps in modeling work and the paper writing. The support of the major project of national science and technology (2000-k01-03-02) is also gratefully acknowledged.

References

[1] Chen S J, Egger J and Han W. Altai-Sayan lee cyclogenesis: A case study. *Beitr. Phys. Atmos.*, 1992, **65**: 13-22.

[2] Han W, Chen S J and Egger J. Altai-Sayan lee cyclogenesis: Numerical simulations. *Meteorl. Atmos. Phys.*, 1995, **55**: 125-134.

[3] Wang S G, Yang D B and Zhou Y S. Discuss on the formation of "94. 4" dust storm on northwest of China. *Chinese Desert*, 1995, **15**: 332-338 (in Chinese).

[4] Hu Q Y. The development of Severe dust storm and dry squall line-An analysis of mechanism on formation of black storm. *Plateau Meteorology*, 1996, **15**: 178-185.

[5] Takemi T. Structure and evolution of a severe squall line over the arid region in northwest China. *Mon. Wea. Rev.*, 1999, **127**: 1301-1309.

[6] Grell G A, Dudhia J and Stauffer D R. *A description of the fifth-generation Penn State/NCAR mesoscale model (MM5)*, 1995, NCAR/TN-398+STR.

[7] Thorpe A J. Diagnosis of balanced vortex structure using potential vorticity. *J. Atmos. Sci*, 1985, **42**: 397-406.

[8] Karyampudi M V, Kaplan M L, Koch S E and Zamora R J. The influence of the Rocky Mountains on the 13-14 April 1986 severe weather outbreak. PartI: Mesoscale lee cyclongenesis and its relationship to severe weather and dust storms. *Mon. Wea. Rev.*, 1995, **123**: 1394-1422.

[9] Wu G X, Cai Y P and Tang X Q. Moisture potential vorticity and slantwise vorticity development. *Acta Meteor. Sinica China*, 1995, **53**: 387-405 (in Chinese).

Wu G X, Cai Y P and Tang X Q and Liu H Z. Complete form of vorticity tendency equation and slantwise vorticity development. *Acta Meteor. Sinica*, 1999, **57**: 1-15 (in Chinese).

Wu G X, Cai Y P and Tang X Q. Comparsion between the complete form vorticity equation and the traditional vorticity equation. *Acta Meteor. Sinica*, 2001, **59**: 384-392 (in Chinese).

[10] Gamo M. Thickness of dry convection and large-scale subsidence above deserts. *Boundary Layer Meteorology*. 1996, **79**: 265-278.

[11] Barenblatt G I, Goltsyn G S. Local structure of mature dust storms. *J. Atmos. Sci.*, 1974, **31**: 1917-1933.

[12] Carlson T N, Benjamin S G and Forbes G S. Elevated mixed layer in the regional severe storm environment: Conceptual model and case studies. *Mon. Wea. Rev.*, 1983, **111**: 1453-1473.

[13] Pauley P M, Baker N L and Barker E H. An observational study of the "interstate 5" dust storm case. *Bull. Amer. Meteor. Soc.*, 1996, **77**: 693-719.

[14] Hacker J P, Mckendry I G and Stull R B. Modeled downward transport of a passive tracer over western north America during an Asian dust event in April 1998. *J. App. Meteor.*, 2001, **42**: 397-406.

[15] Anthes R A, Kuo Y H and Gyakum J R. Numerical simulation of a case of explosive marine cyclogenesis. *Mon. Wea. Rev.*, 1983, **111**: 1174-1188.

[16] Stull R B (Editor), Xu J Q and Yang D R (translator). *An introduction to the boundary layer meteorology*. Qingdao Oceanography College Publishing House, Qingdao China, 1991, pp. 339-342 (in Chinese).

[17] Crespi S N, Artíñano B and Cabal H. Synoptic Classification of the Mixed-Layer Height Evolution. *J. Atmos. Sci.*, 1995, **34**: 1666-1676.

[18] Arritt R W, Wilczak J M and Young G M. Observations and numerical modeling of an elevated mixed layer. *Mon. Wea. Rev.*, 1992, **120**: 2869-2870.

[19] Mahrt L. Mixed layer moisture structure. *Mon. Wea. Rev.*, 1976, **104**: 1403-1407.

导致一例强沙尘暴的若干天气因素的观测和模拟研究*

姜学恭[1]　沈建国[2]　刘景涛[2]　陈受钧[3]

1. 内蒙古自治区气象台,呼和浩特　010051;
2. 内蒙古自治区气象局,呼和浩特　010051;
3. 北京大学物理学院大气科学系,北京　100871

摘　要:利用常规观测资料和数值模拟方法对影响2001年4月6—8日强沙尘暴天气发生的若干天气因素进行了研究。结果表明:在本次"锋面气旋型"沙尘暴过程中,气旋冷锋是引发强沙尘暴的主要系统,冷锋过境对沙尘暴的触发作用远强于气旋发展。冷锋的动力学特点决定其常与强沙尘暴伴随。影响沙尘暴的诸多天气因素中,影响地面大风形成的有效动量下传机制为:高空急流下沉支使高空动量下传至对流层中层,其下方形成的深厚混合层使这一动量继续下传到地面。"混合层"从本质上反映了有利沙尘暴形成的大气层结特征,深厚混合层的是否形成很大程度决定着沙尘暴的强度。本次过程深厚混合层的形成是长时间地面加热的结果,这也是特强沙尘暴仅发生在内蒙古中部偏北地区而不是下垫面条件更为适宜的内蒙古西部的原因。混合层空气的平流对内蒙古中部偏北地区深厚混合层的形成具有相当的贡献,其影响程度与地形分布密切相关。地面热通量试验表明地面加热对冷锋过境时上升运动的强度、深厚混合层形成、高空动量下传等具有重要影响,并因此影响沙尘暴强度。

关键词:沙尘暴,观测研究,数值模拟,混合层

1　引言

1999—2001年,我国北方大部地区频繁受到沙尘天气的侵袭。仅2001年,内蒙古地区即发生21次沙尘暴天气过程,给这些地区的农牧业生产、交通运输、人民生活、生态环境等造成了严重的危害[1]。其中,发生在2001年4月6—8日的沙尘暴是近年来危害较严重的一次沙尘天气过程,其影响波及新疆、甘肃、宁夏、内蒙古、河北、黑龙江、吉林等多个省区,并在内蒙古中部偏北地区引发黑风暴(特强沙尘暴,5个测站能见度小于50 m)。其灾害,仅在内蒙古中部的锡林郭勒盟北部,就造成了包括幼畜死亡30783头,走失3万头,倒塌和损坏棚圈4310间,埋没草场3000亩①。足见其破坏程度之巨[1]。

沙尘暴的形成不仅依赖沙源,同时,大风、热力不稳定等天气因素对其具有重要影响。对于沙源因素,普遍认为沙漠和土地荒漠化地区是主要的供尘区[2,3,4]。全球四大沙尘暴多发区均位于沙漠和荒漠化地区,而我国新疆、青海、甘肃、内蒙古等沙尘暴多发区也多位于沙漠和荒漠化及邻近地区。也有另一些研究[5]认为沙化发展区是主要供尘区,沙尘过程途径地区地表(包括沙化草地、撂荒耕地、退化旱作耕地)所富含的粉尘物质及城市裸土的就地起沙,是沙尘

* 资助课题:中国科技部重点科技攻关项目《沙尘暴成因及综合防治技术开发与示范》。

① 1亩=666.7 m^2,下同。

暴过程的主要沙尘源。

地面大风是沙尘暴形成的必备条件。常引发沙尘暴的天气系统如冷锋、对流系统、气旋、中飑线、陡峭地形强迫[6]等均可产生地面大风。其高空动量下传机制，常由锋后冷平流、飑线对流、地形强迫下沉等过程实现。Pauley 等[7]和 Karyampudi 等[8]研究表明，高空急流出口区次级环流下沉支能够导致对流层高层动量的下传。Hacker 等[9]研究发现，对流层顶折叠引起高空急流下落，再通过湍流垂直混合可以将高空动量有效下传到地面。另外，地形波的诱导也是导致高空动量下传的重要方式。

对沙尘暴形成具有重要影响的另一因素是大气层结条件。大风过程并不总与沙尘暴过程相伴[10]表明还有其他因素影响沙尘暴形成。沙尘暴多发于午后至傍晚，说明热力不稳定对沙尘暴形成具有重要影响[3]。Barenblatt 等[11]研究表明，在稳定层结下沙尘浓度随高度指数减小(假定沙尘粒径相同)，而在对流状态下，沙尘浓度随高度保持不变，进一步说明大气层结状态是沙尘暴强度的重要影响因素。Takemi 等[13]研究发现 1993 年 5 月 5 日黑风暴的形成是由于：地面加热形成的深厚混合层减小了将气块从混合层顶抬升到自由对流高度所需的能量，因而低层的扰动(飑前地面强辐合)足以触发深对流产生，导致强沙尘暴发生。

为了分析地面大风、层结条件等天气因素对沙尘暴发生和强度的影响，本文主要利用常规观测资料并结合数值模拟方法对 2001 年 4 月 6—8 日由蒙古气旋强烈发展引发的沙尘暴过程进行了研究。

2 天气过程概况

本次过程的强沙尘暴天气出现在蒙古气旋强烈发展阶段。5 日 12 时(世界时，下同)前，地面尚未形成锋面气旋(阿尔泰—萨颜岭山地低压)，有扬沙和弱沙尘暴出现在阿尔泰—萨彦岭山地个别测站。5 日 18 时，蒙古气旋生成，气旋冷锋侵入新疆，产生小范围扬沙天气。6 日 00—18 时气旋进入快速发展阶段，冷锋东移侵入内蒙古大部。期间，在 00 时，内蒙古西部阿拉善盟出现大范围扬沙天气。06 时，内蒙古阿拉善盟、巴彦淖尔盟北部、乌兰察布盟北部和锡林郭勒盟西部产生大范围沙尘暴和强沙尘暴，水平能见度下降到 200 m 到 300 m(额济纳旗能见度下降到 0 m)。6 日 12—18 时，气旋发展进入爆发阶段(气压下降 6 hPa/(6 h))，沙尘暴东移，在内蒙古乌兰察布盟北部、锡林郭勒盟达到最强(水平能见度下降到 200 m 以下)。其中，黑风暴发生在 6 日 08 时到 6 日 18 时，出现在内蒙古中部偏北地区二连浩特、满都拉、朱日和、阿巴嘎、苏尼特右旗 5 个测站(能见度 0 m)。7 日 00 时后，气旋形成锢囚并开始减弱，沙尘暴区继续向东扩展，强度减弱。7 日 12 时仅内蒙古东部和黑龙江省少数地区间或有沙尘暴发生。图 1a 为本次过程地面最小能见度分布，可见沙尘暴区呈涡旋带状，强沙尘暴(能见度小于 200 m)主要发生在内蒙古中部，黑风暴出现在内蒙古中部偏北地区。

3 模式、资料和试验设计

采用 Pan State/NCAR 发展的中尺度模式非静力 MM5V3[7]。模式的垂直坐标 $\sigma=(p-p_t)/(p_s-p_t)$，p_s 和 p_t 分别为地面气压和模式层顶气压($p_t=50$ hPa)，垂直方向为 24σ 层。采用三重双向嵌套方案(图 2a，模拟区域)，大网格采用 90 km 水平格距，中心为 45°N，110°E，模拟范围45×41 格点；中网格采用 30 km 格距，模拟范围 73×73 格点；细网格采用 10 km 格距，模拟范围139×

76 格点，分别覆盖气旋中心区和强沙尘暴区。湿过程采用显式水汽方案和 Grell 积云对流参数化方案；行星边界层采用 MRF 高分辨参数化方案；采用云辐射方案和多层土壤模式。

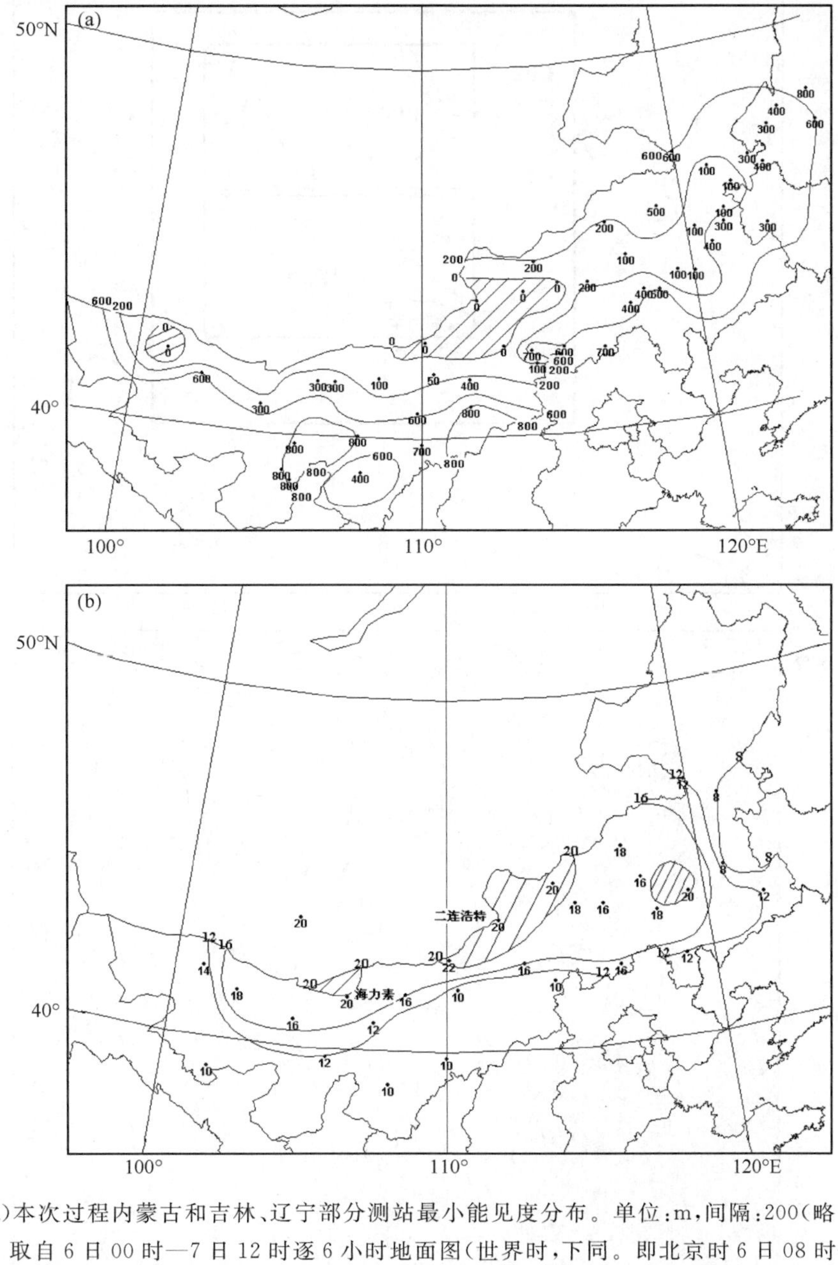

图 1 (a)本次过程内蒙古和吉林、辽宁部分测站最小能见度分布。单位：m，间隔：200（略 400）。斜线为黑风暴区域。取自 6 日 00 时—7 日 12 时逐 6 小时地面图（世界时，下同。即北京时 6 日 08 时—7 日 20 时）；(b)本次过程内蒙古部分测站地面（6 日 00 时—7 日 12 时）风速极值分布。单位：m/s，间隔：4，资料取自逐三小时地面天气图

初始场采用北京有限区域模式（HLAFS）客观分析资料插值得到。地形资料分别采用 NCAR 30 min、10 min、5 min 全球地形资料按不同格距插值生成（图 2b 模拟区域地形）。地表类型采用 NCAR 的 13 类 Landuse 资料。模拟积分时段为北京时 2001 年 4 月 6 日 08 时—2001 年 4 月 8 日 08 时。

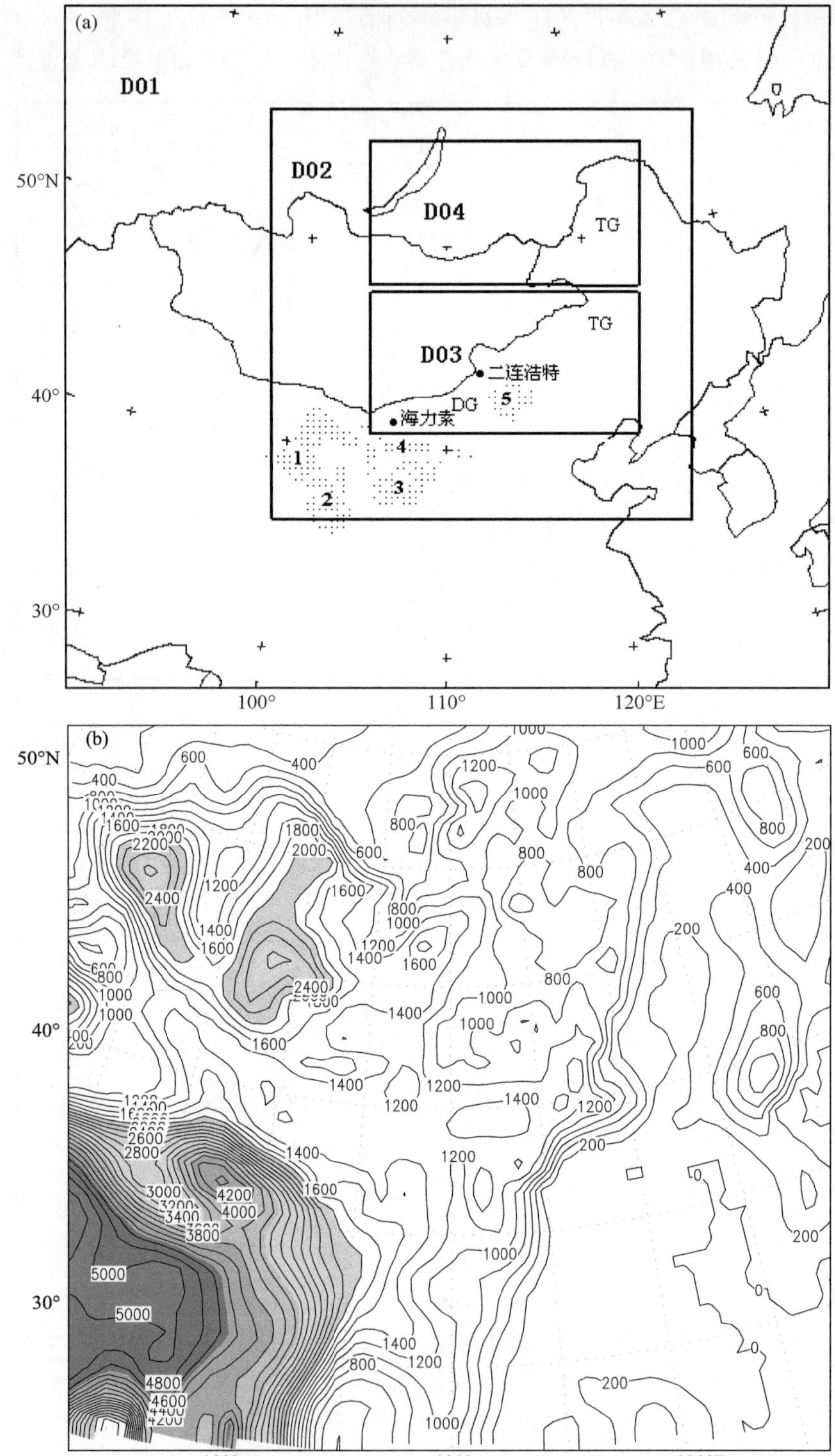

图 2　(a)模拟区域　其中，1. 巴丹吉林沙漠，2. 腾格里沙漠，3. 毛乌素沙地，4. 库布齐沙漠，5. 浑善达克沙地。DG 表示荒漠草原　TG 表示典型草原；(b)模拟区域地形高度，单位：m，间隔：200。阴影区地形高度大 2000 m，北部为阿尔泰—萨颜岭山地

模拟试验方案分别为:(1)对照模拟(Control Experiment,CE):包含以上物理方案和参数选择。(2)地面通量试验(No Surface Flux,NSFE):去掉地面通量并保留以上方案和参数选择。(3)阿尔泰—萨彦岭地形影响试验(No Mountain Experiment,NME):将阿尔泰—萨彦岭山地地形高度取为 1000 m,其他方案和参数选择不变。

4 结果分析

本次过程内蒙古中部沙尘暴强度强于西部(图 1)。为便于比较,选取了表征这两个区域沙尘暴强度的代表站——海力素(一般沙尘暴)和二连浩特(黑风暴)进行分析研究。

4.1 "锋面气旋型"沙尘暴

对于沙尘暴强度较弱的海力素(最小能见度 300 m),从图 3a 可以看到:在锋前(6 日 00 时前),地面风速开始缓慢增大,与之相伴地面气压有所降低(已扣除日变化,图略)。这些变化显然是由气旋发展所致。在此期间能见度由 15 km 降至 8 km,形成扬沙天气。之后冷锋过境(6 日 00—03 时),地面风速迅速增大、风向由西南转为偏西、温度下降、湿度增高。沙尘天气迅速增强,能见度由 8 km 迅速减小到 300 m,形成沙尘暴。可见,在海力素(内蒙古西部),触发锋前扬沙天气的主要原因是气旋发展,冷锋过境则导致沙尘暴的发生。

对于发生黑风暴的二连浩特,在锋前(6 日 09 时前,图 3b),地面气象要素表现明显有别于海力素:其一,增温、减压、减湿现象更为明显(锋前地面 3 小时变压 5.1 hPa;08—14 时地面增温 13℃),其二,风速、风向却非常稳定(锋前为西南风 10 m/s 左右,锋后剧变为西风 16~20 m/s)。因此,这些变化有气旋发展因素,但不排除强地面加热和及强暖平流产生的热对流是另一原因。其造成的沙尘天气明显强于海力素:03 时产生扬沙,06—09 时加强形成沙尘暴(能见度降至 100~200 m)。之后冷锋过境(09 时,北京时 17 时),地面气象要素出现较海力素更为剧烈的变化:气压陡升、温度剧降、风速激增、风向改变、湿度增大(图 3b),表明冷锋更强。相应沙尘暴迅速增强:能见度迅速减小到 100 m 以下,伴随风速继续增大,进一步加强形成黑风暴。因此,在二连浩特(内蒙古中部),锋前沙尘暴源于热对流和气旋发展,冷锋过境使沙尘暴得到迅速增强。

需要指出,首先,本次过程锋前沙尘暴仅出现在午后至傍晚时段,其中强度最强、范围最大在 09 时(北京时 17 时),而气旋发展最强时段(12—18 时,6 hPa/6 h)锋前并未出现沙尘暴,说明锋前沙尘暴产生主要由于热对流等其他因素,气旋发展并不是最重要的。其次,即使在锋前沙尘暴最强阶段(09 时),仅有两个测站出现沙尘暴,这与锋后沙尘暴相比,不论强度(黑风暴出现在锋后)还是范围都明显偏小。因此,通过比较锋面前后沙尘暴强度,并考虑到触发锋前沙尘暴主要是气旋发展因素之外的其他原因,可以得出:对于本例"锋面气旋型"沙尘暴,气旋冷锋是远比气旋发展更为重要的也更为直接沙尘暴触发系统(其形成的涡旋带状的沙尘暴区,也可以推定主要是气旋冷锋圆弧状扫过该区域产生的结果)。

气旋发展常导致地面风速增大并形成低层辐合抬升,同样提供了沙尘天气发生所需的气象条件。它对沙尘暴的触发作用之所以弱于冷锋,结合后文的分析将发现,原因主要是冷锋具有更适宜沙尘暴形成的动力学特点。

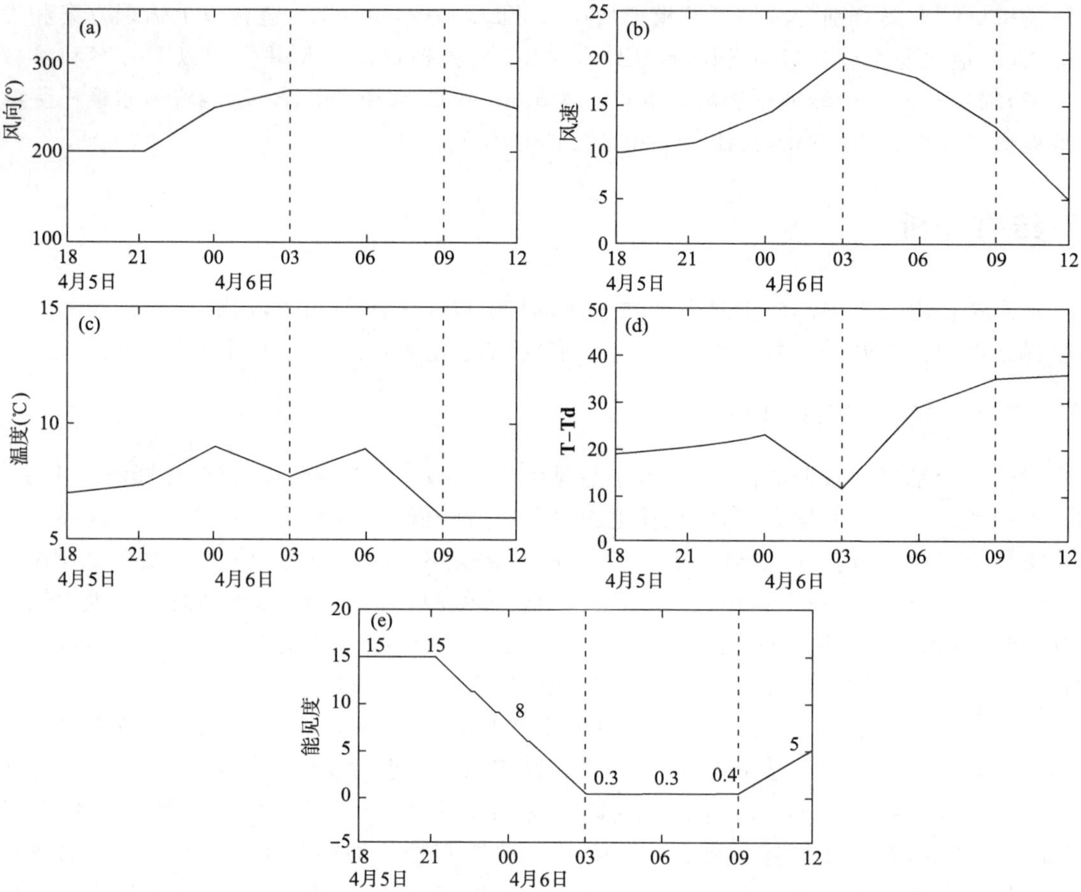

图 3A　海力素 5 日 18 时至 6 日 12 时(北京时 6 日 02 时—7 日 20 时)地面气象要素变化

(a)风向:度;(b)风速:m/s;(c)温度:℃;(d)温度露点差:℃;(e)水平能见度:100 m

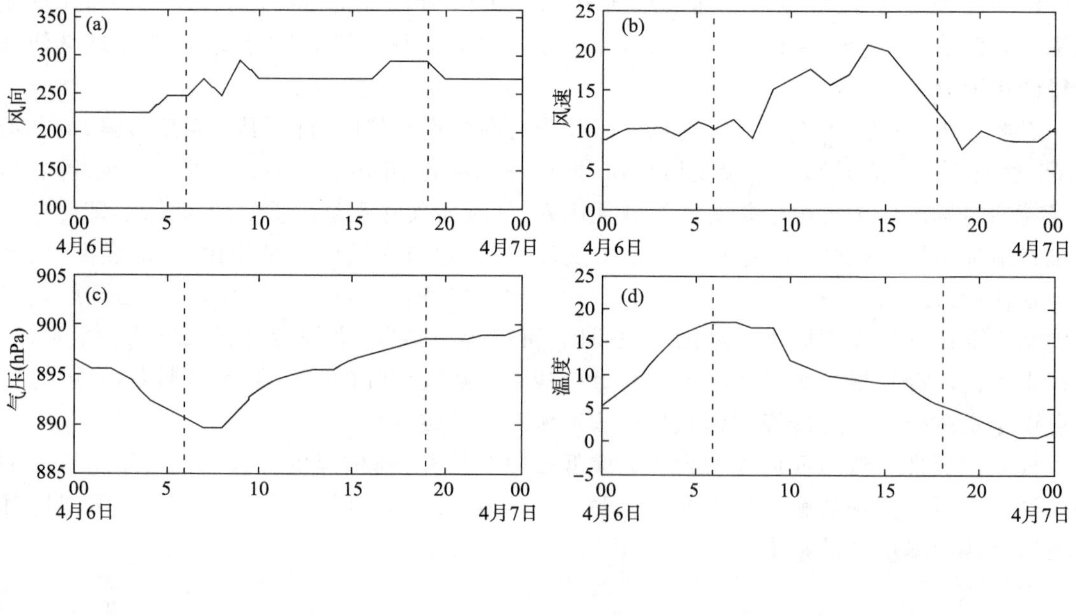

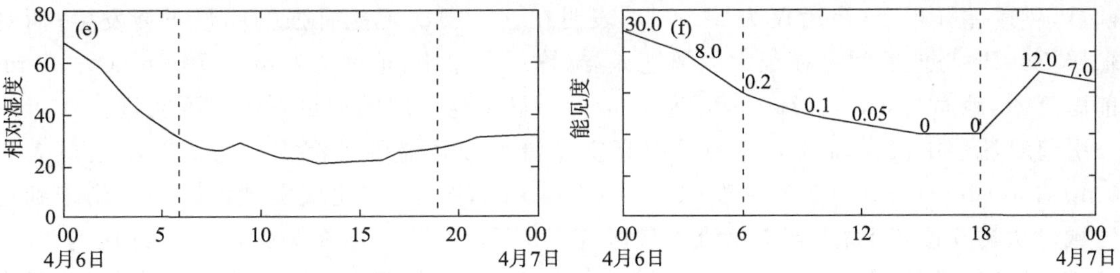

图 3B　二连浩特 6 日 00 时至 7 日 00 时(北京时 6 日 08 时—7 日 08 时)地面气象要素变化

(a)风向:度;(b)风速:m/s;(c)气压:hPa;(d)温度:℃;(e)相对湿度:%;(f)能见度:100 m。虚线表示沙尘暴持续时段

4.2　地面大风

地面大风是影响沙尘暴形成和强度的一个重要因素。本次过程中,在海力素,与锋后沙尘暴同时出现的是地面风速的增大(图 3a,风速从 15 m/s 增至 20 m/s)。在二连浩特(图 3b),当锋后地面风速迅速增大时,能见度从 100～200 m 迅速减小;而与地面风速的两个极值(17 m/s,20 m/s)相对应的地面水平能见度分别是 50 m 和 0 m。可见,沙尘暴的增强与地面风速的跳跃式增加紧密联系。

内蒙古地区过程地面风速极值分布图(图 1b)显示,内蒙古中部的地面风速极值明显强于西部(中部:4 个测站大于 20 m/s,西部:仅海力素为 20 m/s。海力素受地形影响常出现大风天气),这是与能见度分布一致的。下文将分析这两个区域地面大风差异的形成原因。

(1)气旋发展因素

气旋发展是产生地面大风的主要原因(气旋发展通过加大地面气压梯度导致地面风速增强)。本次过程中,地面大风区主要分布在气旋中心附近等压线较密集区(图略),随气旋剧烈发展而增强,随气旋和冷锋的移动向前推进。同时,在冷锋后明显强于锋前(与锋后冷平流有关)。

本次过程中,地面大风的形成也与急流的加强有关。250 hPa 和 700 hPa 天气图(略)显示,气旋强烈发展前(6 日 00 时),高空急流强度较弱(250 hPa,48 m/s),相应低空急流(700 hPa,24 m/s)和地面大风强度也较弱。6 日 06 时后,随着系统东移发展,高空急流迅速增强(12 时:56 m/s),相应低空急流(12 时:32 m/s)和地面大风都迅速增强。可见,地面大风形成也可能受到高空动量下传的影响。

本次过程地面大风也表现出一定的次天气尺度特征。6 日 06—09 时,地面大风区出现明显南伸,20 m/s 大风区扩展到内蒙古中部地区;而在气旋爆发阶段(6 日 12—18 时),地面大风区却明显向气旋中心收缩。说明气旋发展因素之外的其他因素也影响了地面大风形成(地转适应的尺度特点决定气旋发展导致的地面大风应为天气尺度)。以下就导致地面大风的其他因素和高空动量下传机制进行分析。

(2)动量下传

以下的分析仍通过内蒙古西、中部的对比。内蒙古西部海力素附近冷锋过境发生在 00—03 时左右,图 4a 给出了 00 时穿过冷锋的垂直剖面图(此时锋面坡度较小,冷锋强度较弱)。图中风速垂直分布反映对流层中层和低层动量下传存在显著的差异。在 400～700 hPa 间,冷锋前后，风速差异很小(锋后仅为 5 m/s)，表明在这一层次可能有强的动量下传发生；而对流层低层 700 Pa 与地面间存在很大的风速差异(锋后马鬃山：地面为 4 m/s，700 hPa 为 22 m/s；锋前海流图：地面为 4 m/s，700 hPa 为 15 m/s)，显示这一层的动量下传很不明显。

引起对流层中层强动量下传的是由高空急流次级环流的下沉支。通过对高层最大风速(30 m/s，300 hPa)与高空急流强度(48 m/s，250 hPa，图略)的比较发现此时急流尚未到达这一区域。从相互位置上看，高空急流出口区在(43°N，104°E)，而海力素附近的纬度为 41.5°N(106°E)，故这一区域在地面大风形成期间(00—03 时)应位于高空急流出口区右前侧，即次级环流下沉支所在的区域。而高空急流下沉支会产生强的高空动量下传。可见，尽管未进行严格的诊断，但可以基本确定高空急流下沉支是导致内蒙古西部对流层中层强动量下传的主要因素。

需要说明，高空急流与高空锋区、地面冷锋的位置应该是协调一致的(按先后顺序分别是地面冷锋、高空锋区、高空急流)。本次过程之所以出现高空急流下沉支叠置在冷锋后地面大风区之上的现象，是因为此时系统较弱，高空锋区坡度较缓，易于出现这样的配置。

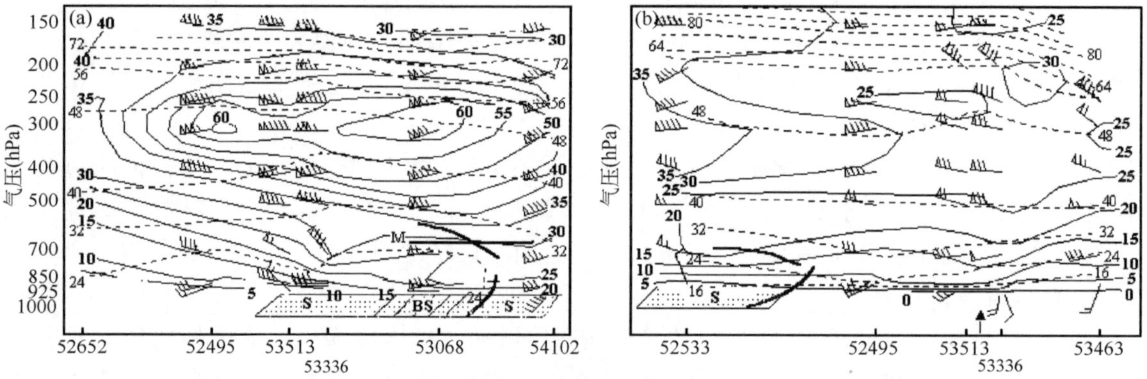

图 4　穿过地面冷锋垂直剖面图

(a)2001 年 4 月 6 日 00 时(箭头为海力素)；(b)2001 年 4 月 6 日 12 时。实线为全风速，单位：m/s，间隔：5；虚线为等位温线，单位：℃，间隔：8。(a)中粗实曲线分别为地面和高空锋区，阴影区为沙尘暴区(BS 为之后出现的黑风暴区)。(b)中粗实线近似表示混合层高度(m)

内蒙古中部的动量下传特点与西部完全相反。这一区域冷锋过境发生在傍晚前后(09—12 时)，图 4b 相应给出 12 时剖面图。值得注意，此时随着气旋在背风坡的迅速发展加强，高空急流(56 m/s，图略)和冷锋的强度都大大增强(图中显示锋面坡度明显变陡)。从高空急流位置看，12 时急流出口区在(43.5°N，117°E)，而二连浩特(09 时冷锋过境)位于 43.8°N，112°E。故 12 时前后二连浩特应位于高空急流的正下方。高层风速对比也说明形成了这一配置：高空急流强度为 64 m/s(250 hPa，图略)，图 4b 中二连浩特上空最大风速在 60 m/s(300 hPa)，两者基本相等。由于位于急流正下方，使得这一区域对流层中层 400～700 hPa 形成了很强的风速垂直切变(30 m/s)。但强的垂直风切变不能产生对流层中层的强动量下传。因为：与海力素对比可见，尽管在二连浩特高空急流远强于海力素(60～30 m/s)，但它们 700 hPa 风速差异并不明显(26～20 m/s)。可见，在对流层中层，高空急流下沉支比急流的加强和下落更能引发高空动量的下传。其内在原因是：高空急流加强和下落是一种大尺度现象，适应过程为风场向温度场调整；而高空急流下沉支的下沉运动是次天气尺度过程，更易形成对风场的调整。其表现形式也不同：急流下落表现为强垂直风切变，而急流下沉支表现为弱垂直风切变。

在对流层中层，内蒙古西、中部动量下传的强弱差异正好抵消了相反的急流强度的差异，使两者未出现 700 hPa 风速的明显差异。因此，造成它们地面大风强度差异的真正原因是对流层低层的动量下传。图 4b 显示：在二连浩特（锋后）700 hPa 与地面风速分别为 30 m/s 和 20 m/s，在锡林浩特（站号：54102，锋前）则分别为 28 m/s 和 18 m/s，风速差异均较弱。说明形成了充分的动量下传，而前文提到海力素附近未形成对流层低层的有效动量下传。因此，真正导致内蒙古中部强地面大风产生的原因是对流层低层的有效动量下传。结合下文的分析可见，造成这一结果的原因是深厚混合层的形成。在二连浩特附近，600 hPa 以下位温差很小（只有一条等位温线），表明形成深厚的混合层；而在海力素附近，600 hPa 以下位温差很大（有三条等位温线），说明未形成明显的混合层。混合层的充分混合显然是有利于动量下传的。

4.3 层结因素

本次过程沙尘暴强度（能见度）分布与地面大风极值并不完全一致（图 1a，b）。在内蒙古西部和东部一些测站都出现了 20 m/s 风速极值，但能见度仅降至 100～300 m，而内蒙古中部部分测站（如锡林浩特）风速虽未达到 20 m/s，但能见度已降至 0 m。可见，影响沙尘暴强度的还包括其他因素。

沙尘暴是一种干对流风暴，因此，干对流强度是沙尘暴强度的一个重要决定因素。能够影响干对流强度的因素，不仅有低层扰动的强弱（如锋面抬升等），还包括有利的层结条件。相关研究指出：热力不稳定层结，即在近地面层形成超绝热层（$\partial\theta/\partial Z<0$），是沙尘暴发生的有利层结条件。但是，这样的超绝热层往往非常浅薄，本例（图 5）中，如以$\partial\theta/\partial Z<0$ 界定，仅 6 日 12 时（二连浩特发生强沙尘暴）出现很弱的超绝热层，高度仅在 850 hPa 以下（12 时前超绝热层应该强于 12 时，但其高度也在 850 hPa 以下，相关研究结果同样如此），而在内蒙古发生沙尘暴的区域，地形高度往往在 1000 m 以上，因此，超绝热层仅仅只有几百米的高度，与经常出现的沙尘粒子 3000 m 左右的扩展高度相比，并不具有非常显著的天气意义。其次，更为重要的是，在干燥环境下，地面加热的影响并不仅仅局限在 850 hPa 以下。由于干旱环境下缺乏充足的水汽以及水汽的平流（沙尘暴发生要求一个干燥的环境，以二连浩特为例，4 月 4—7 日比湿最高值仅为 3 g/kg，沙尘暴发生时则下降到 0.5 g/kg），地面加热将全部用于加热空气，且对空气的加热是以热泡的方式将热量输送至边界层上层，形成具有中性层结特征的混合层，这是干燥环境下行星边界层的显著特点（这不同于湿润环境下具有明显天气意义的不稳定层结，其往往因为有水汽的参与，即形成上干下湿，因此，不稳定层强度很强，也较深厚）。可见，超绝热层不能很好或从本质上反映干旱环境下有利于沙尘暴发生的大气层结特征。

深厚混合层对沙尘暴的意义在于混合层的中性层结有利于深厚对流的产生。同样由于缺乏水汽，干旱环境中往往具有很高的自由对流高度[13]，不利于对流发展。在中性层结层中，抬升气块不需要能量。因此，深厚混合层将有利于深厚对流的产生[13]。Gamo[14]的统计揭示了干旱高原混合层的重要特征：在全球干旱、半干旱高原地区，春、夏季混合层高度常可达到海平面以上 4～6 km。其形成原因是在对流层的中上层（4～6 km）存在一个不发生日变化的中性层（由于缺乏水汽和水汽平流），其下方在夜间为逆温稳定层；在日间为地面加热产生的混合层，两者并未真正连接，但这种类似双不稳定层的层结结构在大气低层出现扰动时很容易产生深厚对流，故垂直混合的层次常达到海平面以上 4～6 km。（海湾战争中浓烟和撒哈拉沙漠的沙尘的扩展高度都达到 500 hPa，被认为是其一个佐证[14]）。

混合层是在近地面行星边界层中形成的中性层结区。对混合层的表征包括混合层高度和强度。作者认为，对于具有天气意义的混合层界定并不需要严格的中性层结。由于缺乏资料，Gamo[14]将其定义为$\partial\theta_v/\partial Z<3.5$ K/km，而 Anthes[15]等曾将自由对流边界层（即日间混合层）定义为$\partial\theta_v/\partial Z<1.5$ K/km，均不影响他们对于相关问题的研究。但其接近中性的程度反映了混合层的强度。对混合层高度的严格确定不能仅仅依靠位温，而需要结合比湿进行判断。这是由于混合层内空气的深度混合往往导致边界层上下层水汽的均匀分布，故比湿垂直梯度往往在混合层顶突然加大，而其下方接近等湿。因此，混合层顶是地面以上位温和比湿梯度急剧增大的高度[16]。

图 5 给出了本次过程 4 月 4—7 日二连浩特的混合层出现的情况（由于水汽含量较小，以位温代替虚位温，从而可以得到 400 hPa 以上层结情况）。结合位温和比湿可见，其一，在沙尘暴发生当日存在明显的混合层日变化：6 日 00 时 700 hPa 以下为夜间辐射逆温层，12 时则形成了深厚的混合层（混合层高度应该在 700 hPa 以上）。其二，其他时间也都存在层结稳定度的日变化，傍晚时稳定度有明显的减小。其三，在 5、7 日 12 时傍晚均有混合层形成，但强度明显弱于 6 日 12 时。其原因可能是云层覆盖及冷平流的影响（4 日 05 时—5 日 20 时云量为 9 成，7 日白天为锋后）。值得注意：在 300 hPa 左右有 Gamo 提到的上部弱层结层存在，但在早晨明显强于傍晚，不同于 Gamo 的结论。另外，图 4 能够大致反映沙尘暴发生时混合层高度的

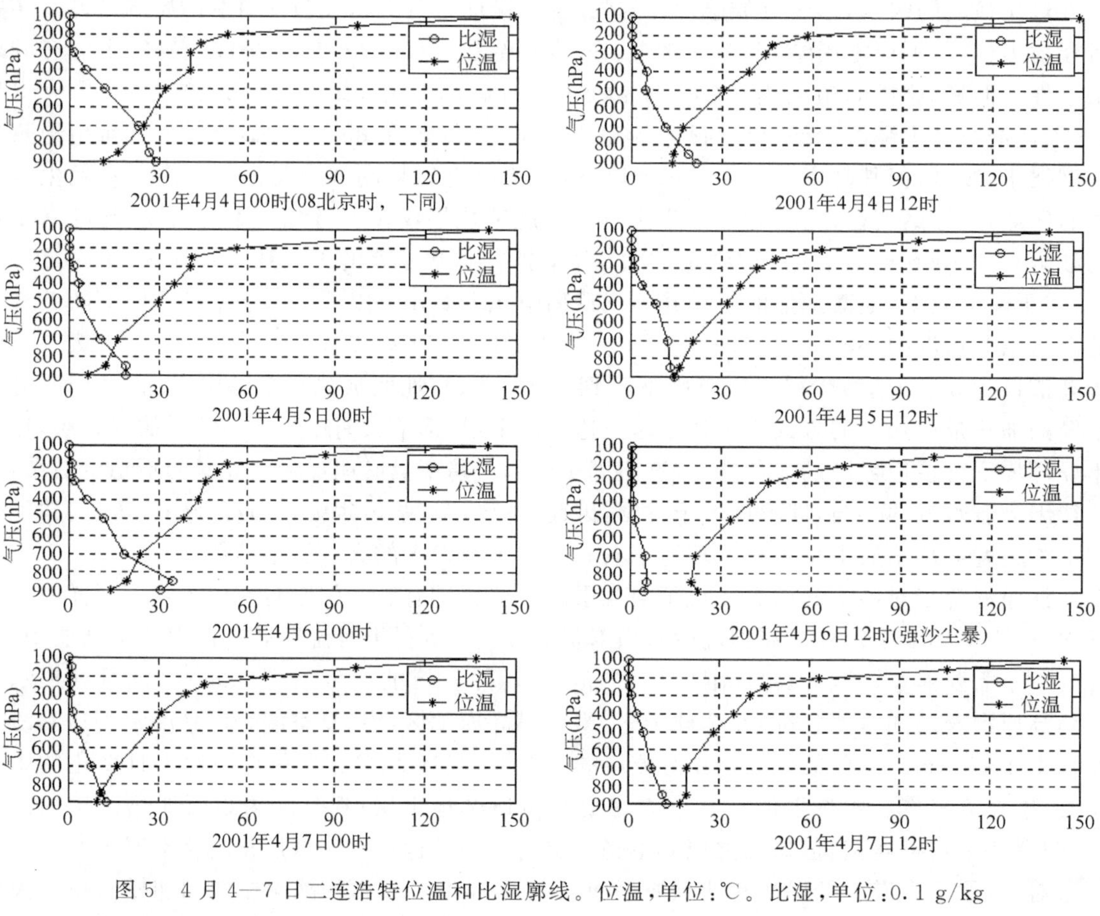

图 5　4 月 4—7 日二连浩特位温和比湿廓线。位温，单位：℃。比湿，单位：0.1 g/kg

空间分布，图中发生黑风暴的内蒙古中部混合层高度明显高于西部：内蒙古中部 600 hPa 以下形成了深厚混合层(位温差很小，仅一条等位温线)；内蒙古西部 600 hPa 以下未形成混合层(位温差很大，有三条等位温线)。可见，混合层对强沙尘暴形成产生了重要影响。

以下的模拟结果可以进一步反映混合层的形成过程及其对冷锋触发干对流产生的影响。图 6 为试验 CE 模拟的相当位温和上升运动时间演变图，图中显示，在二连浩特(图 6a)，6 日 00 时(早晨)开始，地面加热使低层稳定度逐渐减小，逆温层缓慢消失。06 时前后，有混合层形成且在不断抬高，同时有上升运动产生(达 600 hPa 左右，产生锋前沙尘暴)。09 时前后，冷锋过境，锋前混合层达到最高，同时，冷锋的抬升使上升运动迅速增强，并一直扩展至 300 hPa

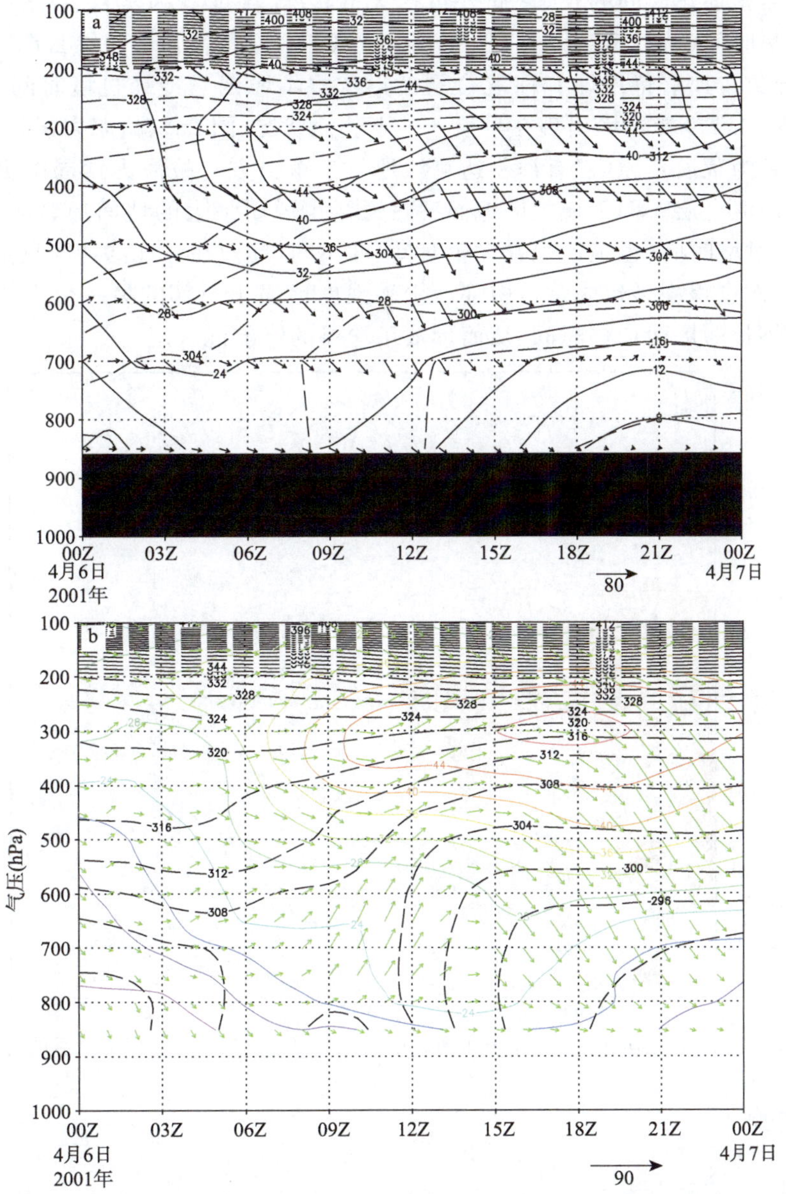

图 6　控制模拟海力素(a)和二连浩特(b)时间—高度剖面图

实线为东西风(U，单位：m/s，间隔：4)，虚线为相当位温(单位：K，间隔：4)，矢量为(U，W 合成，单位分别为 m/s 和 0.001 m/s

（沙尘暴加强至黑风暴）。与之相反，图 6b 显示，在海力素，由于锋前未形成混合层，在 03 时冷锋过境时，上升运动非常微弱（模式初值对上升运动有一定削弱，海力素实际的上升运动要强于模拟结果）。可见，二连浩特锋后深厚混合层的形成，是其强上升运动产生（也是强沙尘暴发生）的一个重要影响因素（敏感试验也将验证这一观点）。

另外，图中在上部中性层（弱稳定层）附近有上升运动产生，并与其下方的上升运动相连接。可见，如果有上部中性层存在，则其对深对流形成具有一定贡献。但内蒙古地区是否存在如 Gamo 所述的上部中性层还需进一步研究。

以上分析可以初步看出影响混合层形成的因素是地面加热。事实上，决定混合层高度和强度的根本因素是加热时间的长短（此外包括天空状况、湿度等因素）。有关研究表明[17]，混合层高度与加热时间（从日出到该时刻）的幂指数成正比（晴空情况下与 ΔT^2 成正比）。就本次过程而言，导致二连浩特深厚锋前混合层的形成根本原因是冷锋过境时间为傍晚（北京时 17 时），相反，海力素冷锋过境出现在早晨。从下垫面状况（图 2a）看，海力素附近有多个沙漠；而二连浩特附近为荒漠草原（仅有浑善达克沙地位于东南方），故海力素周围更易形成较强的地面加热和深厚混合层。图 7 模拟的地面感热通量也说明最大的地面加热区就在这一区域。但是，由于锋面过境时间（也是沙尘暴发生时间）的不同，反而在二连浩特附近形成了深厚的混合层。可见，本次过程中导致内蒙古中部沙尘暴强度远强于西部的另一个重要原因是内蒙古中部获得了足够长的地面加热时间，从而形成了深厚的锋前混合层。

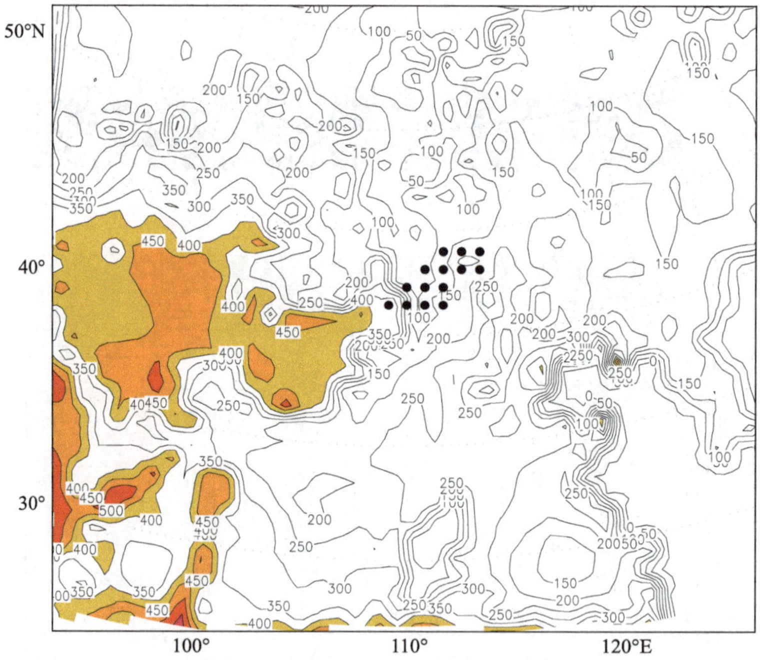

图 7　控制模拟 6 日 14 时地面感热通量分布图。单位：W/m²，间隔：50。点区为之后的特强沙尘暴区域

4.4　敏感试验

（1）地面通量试验（NSFE）

为了进一步分析和验证本次过程地面加热对沙尘暴形成和强度产生的影响，去掉地面热通量进行了 NSFE 试验。

这里，首先需要指出地面加热对气旋冷锋强度也具有重要的影响。图 3b 中二连浩特冷锋过境时表现出非常强的冷锋特征（地面气象要素变化剧烈），而海力素地面气象因素变化很小。仔细比较两站锋面前后地面气象要素变化可见：海力素锋面前后温度大致相当；二连浩特锋后温度与 08 时地面加热前的温度相当。说明地面加热是产生锋面前后温差的重要原因。另外，锋前明显的负变压（−5.1 hPa）和锋后较弱的正变压（2～3 hPa）也说明地面加热可能对冷锋强度有重要影响。另一个需要进一步明确的是在 4.1 小节中指出二连浩特对流层低层之所以形成有效的动量下传，是由于形成了深厚混合层。而地面加热是产生混合层的直接原因，因此，地面加热也是影响对流层低层动量下传的重要因素。

以下 NSFE 试验进一步验证了地面加热的这些影响。对比图 8 与图 6 可见，首先，去掉地面通量导致锋前未形成混合层（变为稳定层结）。另外，上升运动强度大为减小（锋前上升运动强度明显减小；锋后 300～500 hPa 完全为下沉运动）。说明地面加热对混合层形成及深厚干对流触发具有决定性的影响。其次，去掉地面通量导致锋面过境等位温线坡度明显减小，表明冷锋特征变弱（锋后上升运动减弱也表明冷锋减弱）。可见，地面加热是冷锋锋生的一个主要强迫因素。

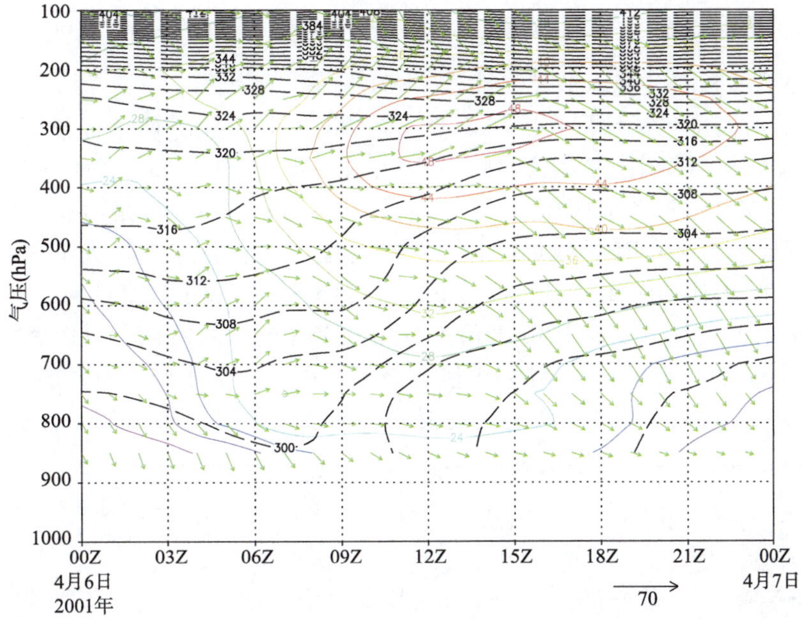

图 8 地面通量试验时间剖面图，类似于图 6

值得注意，图 8 中尽管低层上升运动变弱，但锋后 300 hPa 附近依然有上升运动产生，只是未能与其下方的上升运动形成连接。它的产生显然是在上部中性层存在的情况下，由于低层扰动引起的。因此，如果有上部中性层存在，则对深厚对流形成具有积极的意义。

试验也验证了地面加热对对流层低层动量下传的影响。去掉地面通量使锋前混合层难以形成，这一变化对动量下传的影响可以通过图 9（CE 和 NSFE 的 600 hPa 与 800 hPa 全风速差值）看出，图中，在得益于对流层低层有效动量下传的内蒙古中部（110°E，42°N 附近），CE 的 600 hPa 与 800 hPa 风速差很小（0～2 m/s），反映了之前讨论的这一层次的明显动量下传；而在 NSFE 中，600 hPa 与 800 hPa 风速差很大（8 m/s 左右），说明这一动量下传明显变弱。可见，地面加热是影响本次过程对流层低层有效动量下传的根本因素，也是午后气旋外围大风区出现明显南伸的直接原因。

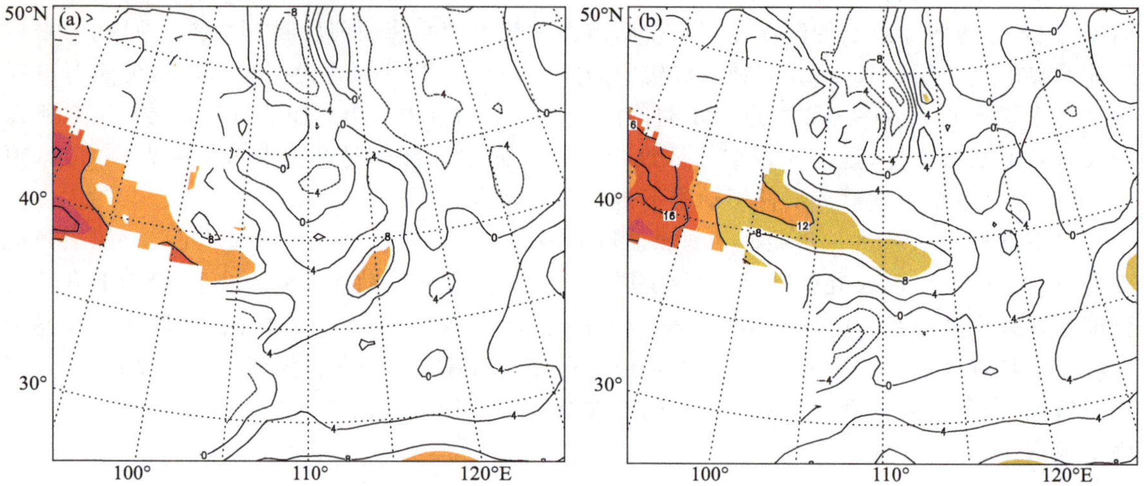

图 9　控制试验(a)和地面通量试验(b)600 hPa 与 800 hPa 全风速差值。单位：m/s，间隔：4。图中阴影区大于 8

(2)混合层平流和阿—萨地形影响试验(NME)

混合层的绝对高度包括地形高度和地形以上高度(相对高度)两个部分。因此，地形和地面加热都将对混合层绝对高度产生影响。这一点在 CE 模拟的 PBL 高度分布图中有清楚的反映，即：PBL 分布是地形和地面加热共同影响的结果(图 10a)。但是，在 NSFE 去掉地面热通量后(只受到地形影响)，PBL 分布(图略)仍然与地形高度分布有很大不同：在阿尔泰—萨彦岭(简称阿—萨)山地东南侧低层气流的下风方(西南侧)形成了较高的与地形高度不一致的 PBL 高值区，即沿低层气流方向 PBL 高值区分布在萨彦岭—蒙古西部—内蒙古中部一线(图 10a 也出现这一现象)。可见，这一结果可能源于混合层空气的平流。有关研究表明[18,19,20]混合层空气可以被平流，从而影响下游地区的混合层高度。典型地形下的混合层平流可以通过

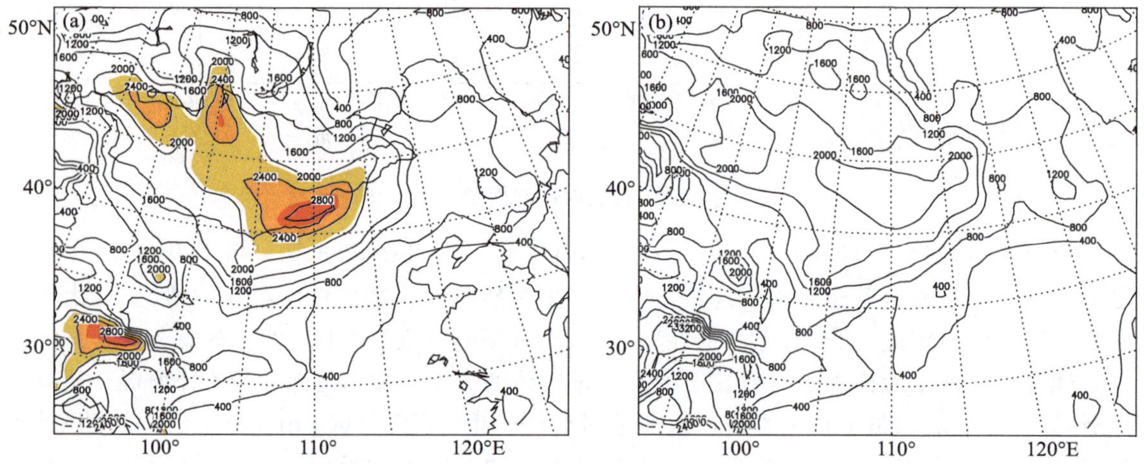

图 10　控制模拟(a)和阿-萨地形影响试验(b)模拟 12 小时 PBL 高度。单位：m，间隔：400。图中阴影区 PBL 高度大于 2100 m，其走向可以反映低层气流对混合层空气平流的方向

图 11 说明。图中，地面加热使山地和平原均产生底部混合层，当气流从高原吹向平原时，高原上空等熵的混合层空气就会被平流到平原上空，形成抬高的混合层(Elevated-ML)，从而使平原上空的混合层高度明显增大。当地形坡度较小时，可不出现双中性层结构而直接使混合层抬高；双中性层结构的天气意义是：使扰动触发的垂直混合易于达到 E-ML 高度。

为了验证本次过程平流对混合层高度的影响，试验 NME 将阿尔泰—萨彦岭山地地形高度减至 1000 m(与周围地形高度一致)进行了模拟，试验 CE 沿气流方向的剖面图(图略)反映出平流作用使山地下游的混合层明显抬高；而 NME 中，由于山地被削平，平流未导致山地下游混合层高度发生明显变化。对比 CE 和 NME 的 PBL 高度分布(图 10a、b)可见，在内蒙古中部偏北地区黑风暴发生区域，PBL 高度从 CE 中的 2800 m 减至 NME 中的 2000 m。可见，平流作用是导致该地区锋后深厚混合层形成的另一个重要因素(由于西北气流仅出现在冷锋后)。

一些相关的统计研究结论可以一定程度地支持混合层平流的观点。由于地面加热而形成的日间混合层往往在地面加热减弱、停止时(即黄昏)达到最高，但许多统计结果表明，强沙尘暴除出现在午后之外也更多出现在前半夜。其直接原因可能就是与强沙尘暴区伴随的深厚混合层的移动(平流)。另外，需要指出，混合层平流并不局限在存在明显地形差异的环境下，只要存在上下游混合层高度差，平面地形情况下仍可产生混合层的平流。而地形对平流的影响在于：图 11 中的低层气流方向与地形高度分布的适宜配置，会明显增强混合层平流对下游的影响程度，这也是本次过程混合层的平流影响之所以集中在锋后的原因(西北高东南低的地形特点决定锋后西北气流能够产生更大的平流贡献)。

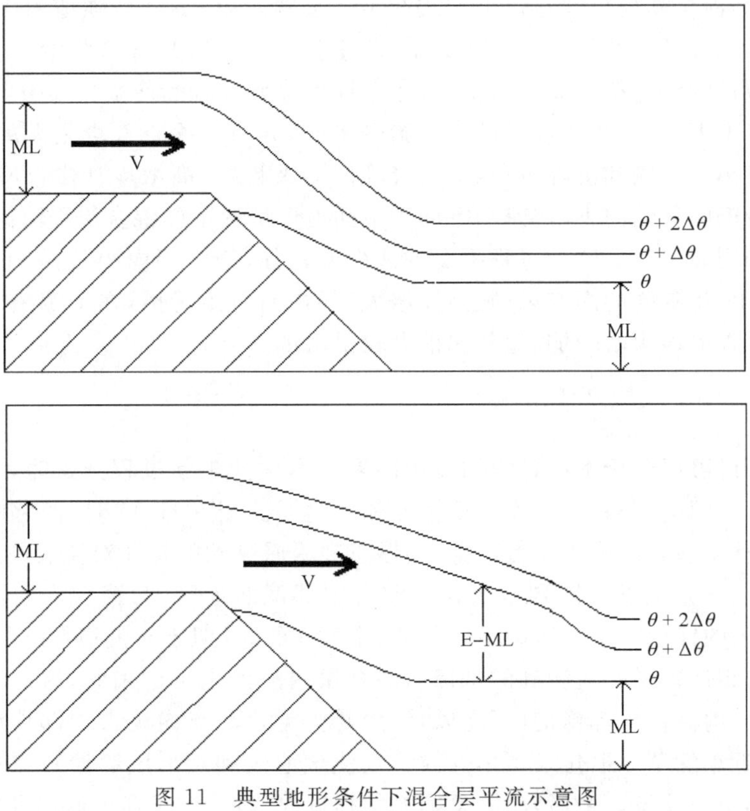

图 11 典型地形条件下混合层平流示意图

4.5 “冷锋型”沙尘暴概念模型

4.2小节主要讨论了混合层层结特点对触发强对流的贡献，混合层的另一个重要特点是它能够反映沙尘暴发生的动力学本质。Barenblatt 等[11]指出：稳定层结中沙尘浓度随高度指数减小；对流状态下沙尘浓度随高度保持不变。这一观点揭示了沙尘暴发生时强干对流产生的充分混合导致沙尘浓度上下一致这一动力学本质（产生沙尘暴必须完成从起沙到沉降整个过程）。强沙尘暴形成的直立高耸的沙尘墙则直观地体现了这一特征：充分混合及沙尘浓度的均匀分布。混合层的中性层结状态和充分混合的特征满足了这一动力学本质的要求，即：充分混合使混合层中的沙尘粒子均匀的分布。

尽管混合层能够体现沙尘暴发生的动力学本质，但是，不是所有情况下出现的混合层都能与（强）沙尘暴伴随，因为沙尘暴的产生，还要求沙尘具有足够大的粒径和足够高的浓度。对锋面系统而言，能够达到这一要求的混合层出现在冷锋后。这是因为，首先，冷锋后往往形成深厚的冷空气堆（无地面加热也可形成，可参见本文图8和其他文献），冷空气堆内的层结是等熵的（其上方为稳定层结），这是锋前抬升和锋后下沉产生的充分混合造成的，相当于形成混合层，冷锋越强，冷空气堆越深厚（即锋后混合层，图6）。其次，冷锋过境产生的强地面风和强抬升能够产生大粒径和高浓度的起沙和扬沙（对比本次过程锋面前后沙尘暴强度，可说明这一点）。所以，能够体现其动力学本质并满足其物质要求（沙尘）因而能够与（强）沙尘暴伴随的是锋后冷空气堆（锋后混合层）。

因此，冷锋型沙尘暴应该通过以下方式产生。起沙阶段：冷锋后强地面大风产生起沙（足够多的且包含大粒径的沙尘）。扬沙阶段：冷锋过境产生强抬升将这些沙尘上扬到对流层中层，其中，较大粒径的沙尘抬升高度较低，分布于锋后等熵混合的冷空气堆中（假设局部沙尘粒径相同），其浓度上下一致；较小粒径的沙尘抬升较高，分布于冷空气堆之上的稳定层中，其浓度随高度指数减小。传输和沉降阶段：冷空气堆内的大粒径、高浓度沙尘，以混合层整体向前推进的方式随冷锋的移动向下游传播，等同于平面地形情况下的混合层平流，其充分混合的特征使沉降与传输同时发生。这一过程引发的沙尘暴较强；稳定层内小粒径、低浓度（浓度随高度指数减小）沙尘，将随强的高空风（随高度增大）很快向下游传播，在下游由下沉运动诱发沉降。这一过程产生的沙尘暴较弱，但其比前者较早出现。

4.6 沙源

基于以上探讨可以分析本次过程的沙源因素。本次沙尘暴过程开始时，首先在阿尔泰—萨彦岭山地及蒙古国南部Gebi沙漠出现扬沙和沙尘暴。至6日06时，遥感图（略）上有大面积质地均匀、平整的沙尘区位于蒙古南部，其形态表明遥测到的是在对流层中层（稳定层）向下游传输的沙尘（其下方难以探测到）；而在内蒙古中部偏北地区，与锋后表面均匀的沙尘区相比，二连浩特附近的锋前沙尘暴区为不均匀的、小范围的、类似于对流泡形态的沙尘区（与中尺度热对流的特征相似），说明遥测到的是锋前混合层内的沙尘暴。因此，这一区域和时段沙尘来源应该有两个：内蒙古中部锋前（西南风）沙尘暴应主要源于内蒙古中部荒漠草原的局地起沙（内蒙古中部近年的草原退化及干旱气候背景使该地区利于形成起沙）；而在高层传输的沙尘应来自蒙古国南部沙漠。之后，冷锋呈圆弧状扫过内蒙古，由于锋前起沙和锋后（混合层内）沉降是相伴发生的，故内蒙古地区荒漠草原区的局地起沙是强沙尘暴的一个主要沙源，来自蒙古国南部沙尘的远距离输送是沙源的另一组成部分。之后,气旋冷锋移动到内蒙古东部并形成锢囚,而这一区域为典型草原区,故局地起沙较弱,沙尘应来源于内蒙古中部的起沙和远距离传输(蒙古国南部已不重要),此时(7 日 00 时)产生的沙尘暴较弱。

值得指出:7 日 00 时遥感图上在内蒙古东部和东北地区为涡旋状的、大范围的、表面均匀的沙尘区覆盖,直观看其强度远大于内蒙古中部出现的沙尘暴,但地面能见度并不很低(300 m 以上)。说明遥感监测到的表面均匀的沙尘区是在对流层中层(稳定层)远距离传输的沙尘,遥感监测的沙尘暴强弱并不能反映实际的地面沙尘暴强度,即我们日常关注的地面能见度。

我国北方的地理和生态特点一定程度加剧了沙尘暴向东部地区的扩展程度。首先,由于强沙尘暴区是以混合层平流的方式向前推进,因此,地形分布和低层气流方向对其具有重要影响。我国北方西高东低,常发生沙尘暴的干旱、半干旱地区又位于西部高原,因此,当盛行西北气流的春季来临时,地形影响使高原地区形成的强沙尘暴极易随西北气流向其东部的平原地区扩展。其次,高原上经常出现的深厚混合层,使沙尘很容易被抬升到足以实现远距离传输的高度,也导致我国东部及亚洲其他地区容易受到西部高原地区的沙尘天气影响。这是特定地区的地理、生态因素产生的不可完全抗拒的结果。

5 结论

本文针对一例由蒙古气旋强烈发展引发的特强沙尘暴过程进行了观测研究和数值模拟分析,结果发现:

(1)在"锋面气旋型"沙尘暴过程中,气旋冷锋是更为重要和直接的沙尘暴触发系统,并常与强沙尘暴伴随。在本次过程影响沙尘暴的天气因素中,地面大风的形成主要受到气旋发展和锋后冷平流的影响;高空动量的有效下传是另一重要原因。不同区域导致地面大风的动量下传机制不尽相同:在内蒙古西部,高空急流下沉支完成将动量从对流层高层到中层的输送,在低层由于未形成深厚混合层使动量下传并不明显;在内蒙古中部,高空急流的加强和下落使高空动量下传到对流层中层,而深厚混合层的形成使这一动量能够继续下传到达地面。这一结果也说明,高空急流下沉支和混合层内的深度混合分别是对流层中层和低层动量下传的有效方式。

(2)不同于降水过程中的不稳定层结,由于地面加热形成的深厚混合层是干旱环境下一个最根本的层结特征。同时,混合层也很好体现了有利沙尘暴发生的大气层结特征。强和特强沙尘暴发生在内蒙古中部偏北地区,根本原因是长时间地面加热形成了深厚混合层。其有利于深厚对流产生的特点,导致冷锋过境触发深厚干对流并引发强沙尘暴。

(3)阿尔泰—萨颜岭上空深厚的混合层空气被低层气流平流是导致内蒙古中部偏北地区锋后深厚混合层形成的另一重要原因。混合层空气的平流与地形高度分布密切相关。

(4)地面通量试验表明:地面加热对冷锋抬升产生的上升运动具有非常重要的影响,是本次过程冷锋锋生的主要强迫机制。同时,地面加热也直接影响混合层形成进而影响对流层低层的动量下传,并通过它们影响沙尘暴强度。

(5)本次过程的沙尘主要来源于蒙古国南部沙漠地区的远距离沙尘输送和我国内蒙古中部荒漠草原区的局地起沙。

(6)混合层对沙尘暴过程具有的另一重要意义是体现了沙尘暴发生时的动力学本质:即充分混合导致沙尘均匀分布,混合层高度和强弱一定程度决定着沙尘暴的强度。冷锋后冷空气堆既体现了(强)沙尘暴的这一动力学本质,也满足了其要求的物质条件(大粒径和高浓度沙尘),因而常与强沙尘暴伴随。冷锋型沙尘暴过程中,沙尘被扬起后,主要通过两种方式产生影响,其一:大粒径、高浓度沙尘在锋后充分混合的冷空气堆中整体向前推进(传输和沉降同时发生),引发强沙尘暴;其二,小粒径、低浓度沙尘在对流层中层随高空风快速向下游传播并沉降,产生弱沙尘天气。

(7)我国北方地理和生态特点一定程度加剧了高原地区沙尘暴的影响。

参考文献

[1] 刘景涛等:蒙古气旋爆发性发展形成强沙尘暴的个例研究.气候与环境研究.2003,(2)待发表.

[2] 王式功,扬德保等.我国西北地区黑风暴的成因和对策.中国沙漠,1995,**15**:19-30.

[3] 王式功.我国西北地区沙尘暴时空分布及成因分析.中国科协技术年会论文集,北京,中国科协出版社.1995.

[4] 刘景涛、郑明倩.华北北部黑风暴的气候学特征.气象,1998,**24**(2):39-44.

[5] 叶笃正,丑纪范等.关于我国华北沙尘天气的成因与治理对策.地理学报,2000,**55**:513-521.

[6] Brazel A J. The Relationship of weather types to dust storm generation in Arizona(1965—1980). *Journal of Climatology*,1986,**6**:255-275.

[7] Pauley P M,Baker N L and Barker E H. An observational study of the"interstate 5"dust storm case. *Bull. Amer. Meteor. Soc*,1996,**77**:693-719.

[8] Karyampudi M V,Kaplan M L,Koch S E and Zamora R J. The influence of the Rocky Mountains on the 13—14 April 1986 severe weather outbreak. PartI:Mesoscale lee cyclongenesis and its relationship to severe weather and dust storms. *Mon. Wea. Rev.* ,1995,**123**:1394-1422.

[9] Hacker J P,Mckendry I G and Stull R B. Modeled downward transport of a passive tracer over western north America during an Asian dust event in April 1998. *J. App. Meteor.* ,2001,**42**:397-406.

[10] 瞿章等."93.5.5"沙尘暴的若干启示,干旱与地理,1994,**17**:63-67.

[11] Barenblatt G I,Goltsyn G S. Local structure of mature dust storms. *J. Atmos. Sci.* ,1974,**31**:1917-1933.

[12] Anthes R A,and Warner T T. Development of hydrodynamics models for air pollution and mesometeorological studies. *Mon. Wea. Rev*,1978,**106**:1045-1078.

[13] Takemi T. Structure and evolution of a severe squall line over the arid region in northwest China. *Mon. Wea. Rev.* ,1999,**127**:1301-1309.

[14] Gamo M. Thickness of dry convection and large-scale subsidence above deserts. *Boundary Layer Meteorology*. 1996,**79**:265-278.

[15] Anthes R A,and Warner T T. Development of hydrodynamics models for air pollution and mesometeorological studies. *Mon. Wea. Rev*,1978,**106**:1045-1078.

[16] Mahrt L. Mixed layer moisture structure. *Mon. Wea. Rev.* ,1976,**104**:1403-1407.

[17] Crespi S N,Artíñano B and Cabal H. Synoptic Classification of the Mixed-Layer Height *Evolution. J. Atmos. Sci.* ,1995,**34**:1666-1676.

[18] Carlson T N,Benjamin S G and Forbes G S. Elevated mixed layer in the regional severe storm environment:Conceptual model and case studies. *Mon. Wea. Rev.* ,1983,**111**:1453-1473.

[19] Stull,R B 著徐静琦、扬殿荣 译.边界层气象学导论.青岛:青岛海洋大学出版社.1991,457pp.

[20] Arritt R W,Wilczak J M and Young G M. Observations and numerical modeling of an elevated mixed layer. *Mon. Wea. Rev.* ,1992,**120**:2869-2870.

Observational and Numerical Study of Some Weather Factors Assciating with a Severe Dust storm

JIANG Xuegong[1] SHEN Jianguo[2] LIU Jingtao[1] CHEN Shoujun[3]

1. Inner Mongolia Meteorological Observatory, Huhhot, China;
2. Inner Mongolia Meteorological Bureau, Huhhot, China;
3. Department of Atmospheric Sciebces, Peking University, Beijing China

Abstract Observational study and numerical simulation were conducted on a severe dust storm process induced by the intensive development of Mongolia cyclone occurred from 6 to 8 April 2001 in Inner Mongolia. The results shown that in this "cyclonic dust storm" process the cyclonic cold front was the direct and important portion of cyclone system inducing the dust storm. The strong surface wind, which associated with the strength of dust storm closely, was mainly caused by the rapid cyclogenesis and the cold advection. And the reason of the maximal surface wind formed in the middle-north part of Inner Mongolia was the formation of deep daytime mixed layer (ML), which transported the great momentum in the mid-lower troposphere down to the surface.

The ML was the crucial stratification factor affecting the formation of dust storm. It's depth and strength mostly determined the strength of dust storm. In this case the reason that the black storm occurred in the middle-north part of Inner Mongolia was the formation of deep ML, which was induced by the surface heating for a long time. The advection of the ML air also contributed to the formation of deep ML there. Then, the deep dry convection was induced when the cold front passed and the severe dust storm occurred.

The sensitivity experiment proved that the surface heating was the crucial factor inducing the formation of deep ML. It also influenced the strength of surface wind by affecting the downward momentum transportation in the lower troposphere. Through them the surface heating worked on the formation of dust storm.

Key words: Dust storm, Observational study, Numerical simulation, Mixed layer

蒙古气旋(低压)的若干统计特征*

姜学恭[1]　康　玲[1]　胡英华[1]　马清霞[2]　李海英[2]

1. 内蒙古自治区气象台,呼和浩特　010051;2. 内蒙古自治区气候中心,呼和浩特　010051

摘　要:利用1948—2004年NCEP海平面气压场再分析资料对蒙古气旋(低压)的月际、年际、年代际变化和强度、位置特点及其与气候因子的关系进行了统计分析。结果表明:蒙古气旋(低压)发生频数的高峰出现在20世纪60年代末70年代初,之后呈现减少趋势。蒙古气旋(低压)主要出现在春季,其次依次为秋、冬季,其中5月最多。中心气压为(1000～1010 hPa)的蒙古气旋(低压)出现次数最多,小于990 hPa的相对较少。蒙古气旋(低压)主要出现在阿尔泰—萨彦岭山地背风坡,其次为46°N以南。蒙古气旋(低压)频数距平与西风环流指数大致呈现反位相分布,表明中纬度经向型环流更利于蒙古气旋产生。这些结果可以为蒙古气旋的预报、预测提供参考。

关键词:蒙古气旋(低压),气候特征,统计分析

引言

由于受到阿尔泰—萨彦岭山地地形影响,在蒙古境内阿尔泰—萨彦岭山地背风坡一侧,往往较为频繁地出现地面低值系统。这些系统一般表现为两种形态,在冷空气较为活跃和有较强冷暖空气交汇的季节(如春季),常形成蒙古气旋[1],而在冷空气势力较弱时,常常形成地面低压。地面低压往往强度较弱,维持时间很短,对天气的影响不大。而蒙古气旋是影响我国的重要天气系统,尤其在春季,不仅发生频率高,而且影响范围广,常常波及蒙古国及我国华北,东北甚至华东地区。蒙古气旋常常导致多种灾害性天气,如大风、寒潮、强降雪等[2]。特别是在春季,蒙古气旋是导致我国华北地区沙尘暴的主要天气系统,如近几年影响我国北方的沙尘暴过程中,大多数都是由蒙古气旋引发的[2,3]。目前国内对蒙古气旋(低压)特征的研究还不是很多。因此,有必要利用长序列的NCEP再分析资料,对蒙古气旋(低压)的气候规律进行分析研究。

1　使用资料

使用资料为1948—2004年NCEP日平均海平面气压场再分析资料。资料分辨率为2.5°×2.5°。使用日平均而不是每日四次的海平面气压场资料,是因为这样可以保证统计的气旋(低压)具有一定的强度。由于形成于阿尔泰—萨彦岭山地背风坡的许多低压往往很弱,且维持时间很短,这些系统不会对天气产生重要的影响,因此,利用日平均资料可以一定程度地减小这些低压对结果的影响。

统计方法是通过计算机程序判断在(42.5°—50°N、102.5°—120°E)范围内是否出现闭合低压(两条等值线,5 hPa)。如果出现,则表明出现一次蒙古气旋(低压)日数。为了验证该统

* 资助课题:国家自然科学基金(40565003)。

计方法的准确性,通过与2001年3—5月和2003年3—5月地面天气图蒙古气旋(低压)的出现次数进行了对比,两者吻合率达到80%以上,表明该统计方法是可行的。

2 结果分析

2.1 蒙古气旋(低压)发生频数的气候特征

2.1.1 年变化

图1为蒙古气旋(低压)1948—2004年(2004为半年)发生频数的年际变化,图中可以看到蒙古气旋(低压)年出现次数在70次左右。其中,从20世纪50年代末到60年代初和60年代末到70年代为两个明显的峰值,均超过100次,之后总体呈现下降趋势,而目前蒙古气旋(低压)的出现次数基本上维持在平均略偏少的水平。

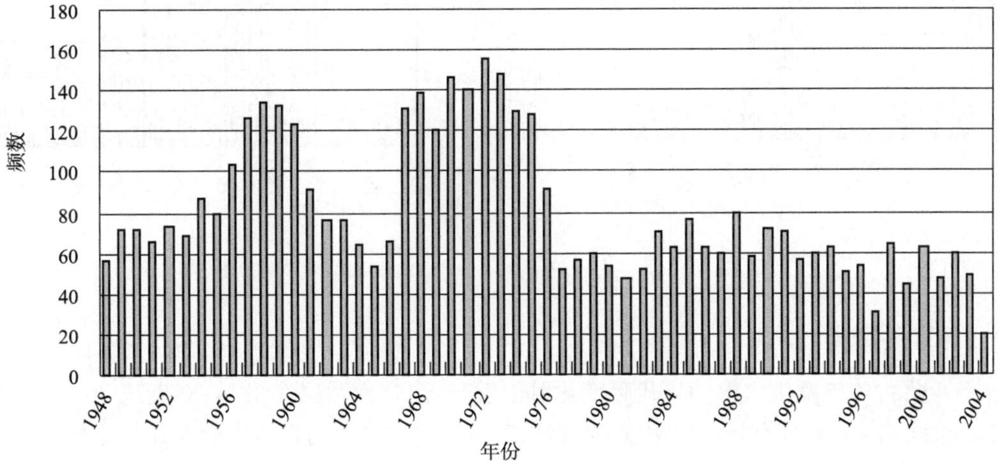

图1 1948年—2004年蒙古气旋(低压)发生频数

2.1.2 月变化

蒙古气旋多发生在春季,统计结果验证了这一事实。图2为1948—2004年累计各月蒙古气旋发生频数变化,图中可见,5月蒙古气旋发生次数最多,之后依次为4月、10月、3月,出现次数最少为12月和1月。可见,春秋季蒙古气旋(低压)的发生次数远大于其他季节,深层原因是在春、秋季节冷暖空气交汇频繁,导致蒙古气旋的出现次数较多,而在冷空气占主导地位的冬季,蒙古气旋的出现次数则相对较少。

前文结果显示,冬季蒙古气旋(低压)的出现次数远小于春季,图3中1月和4月的年变化也说明这一点。另外可以看到,在春季的4月,蒙古气旋(低压)发生次数的年变化并不明显,尽管发生次数最多的年份也出现在20世纪60—70年代,但差异不大,而在冬季,平均的蒙古气旋(低压)发生次数很少,不足5次,但在60年代末到70年代中期却出现一个明显的高峰期。在其他月中,2月、3月、11月、12月与1月类似,而5月与4月类似。因此,图1中蒙古气旋(低压)在60—70年代出现的高峰期主要是由于秋冬季节出现次数增多导致的,春季蒙古气旋各年的发生次数变化并不明显。

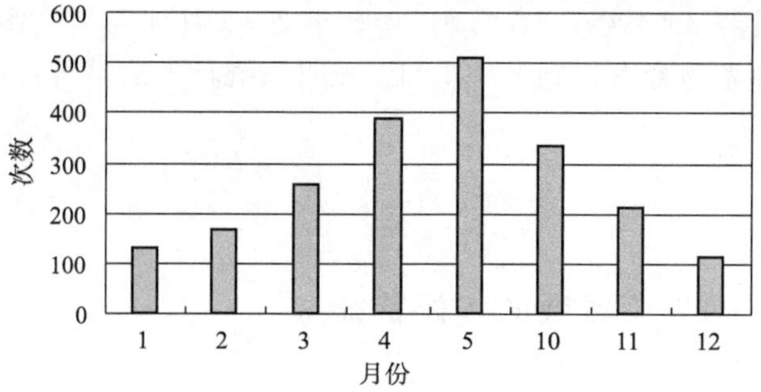

图 2 1948 年—2004 年各月累计蒙古气旋(低压)发生次数

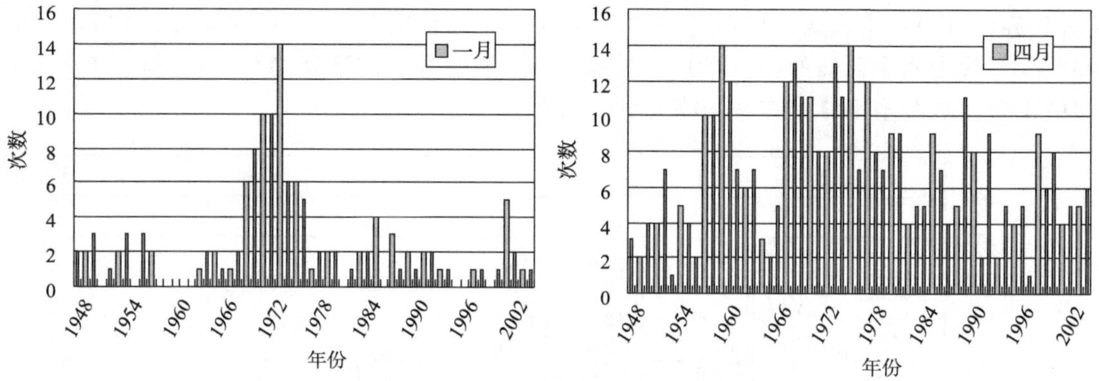

图 3 1 月和 4 月蒙古气旋(低压)发生次数年纪变化

2.1.3 年代际变化

从图 4 蒙古气旋发生次数的年代际变化可以看到,出现次数最多的是 20 世纪 70 年代,然后依次为 60、50、80 年代,90 年代出现次数最少,目前蒙古气旋(低压)的发生次数明显处于较少的阶段。

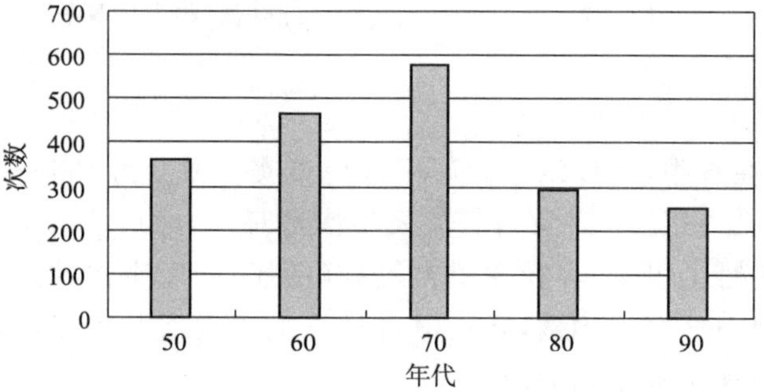

图 4 蒙古气旋(低压)发生次数的年代际变化

2.2 蒙古气旋(低压)的强度特征

图 5 为 1948—2004 年蒙古气旋(低压)强度分布,图中可见,蒙古气旋的强度分布基本上接近于正态,中等强度(1000~1010 hPa)的气旋(低压)出现的次数最多,其次为稍弱的1010~1020 hPa,再其次为 990~1000 hPa,非常强的(980~990 hPa)和非常弱(大于 1030 hPa)的蒙古气旋均出现次数较少。能够对天气产生较大影响的气旋强度一般都需要达到 990 hPa 以

下,可见,这类气旋在蒙古气旋(低压)中所占比重并不大。

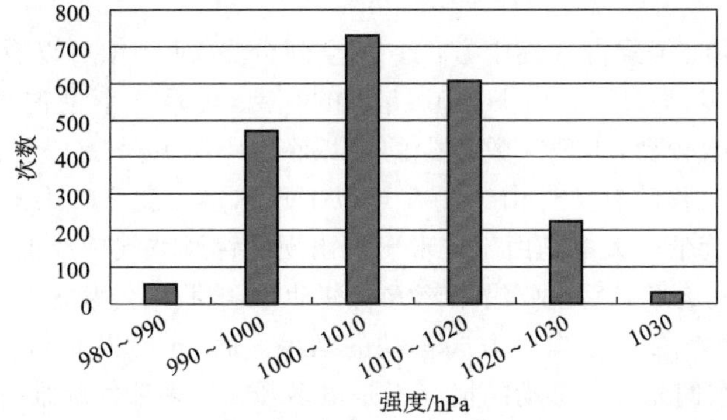

图 5　1948—2004 年蒙古气旋(低压)强度分布特征

图 6 为 1948—2004 年不同强度蒙古气旋的发生频数变化,图中可见,较弱的蒙古气旋(低压)(1000～1010 hPa)各年出现次数变化不大,在 90 年代后期有增多的趋势。而强(990～1000 hPa)和特强(980～990 hPa)的主要出现在 20 世纪 60—70 年代,之后迅速减少。特别是(980～990 hPa)的特强蒙古气旋,在 80 年代后期到 90 年代很少发生,只在 90 年代末到本世纪初有明显增强的趋势。

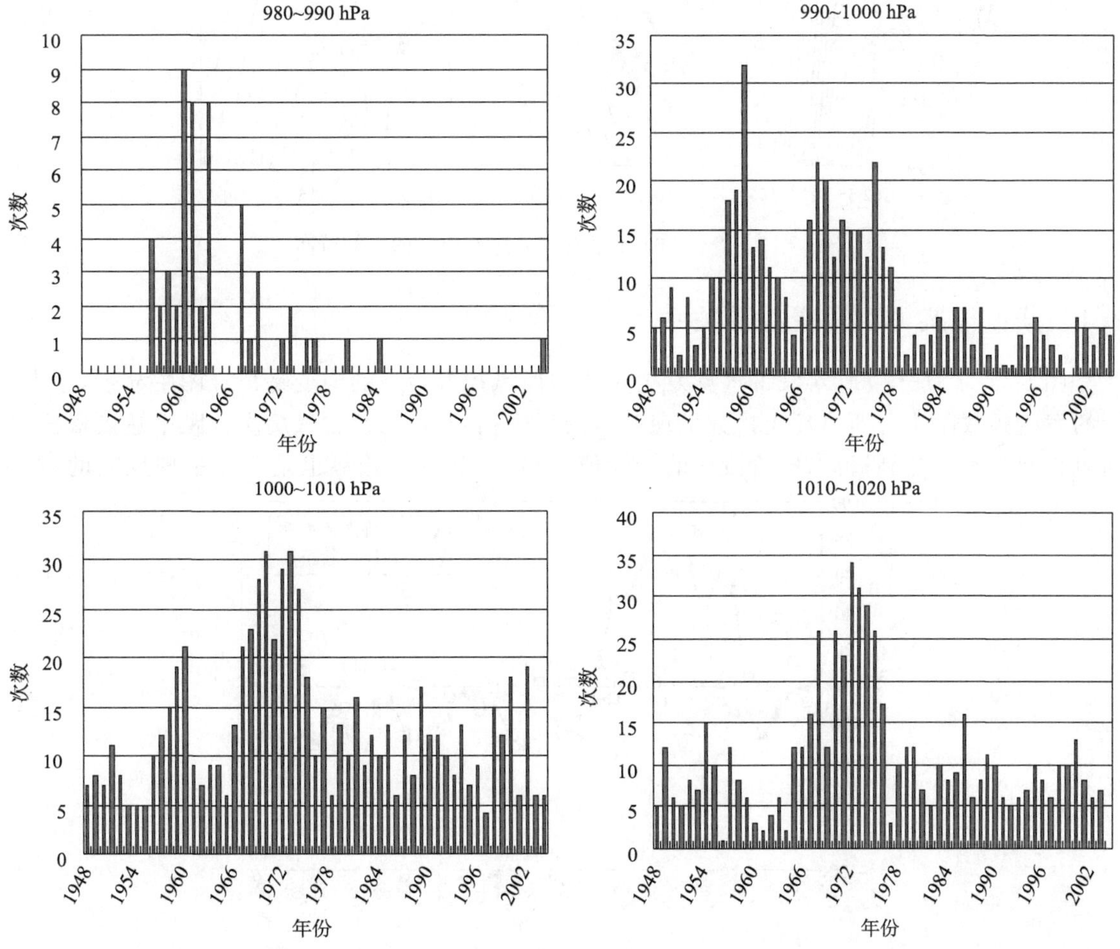

图 6　1948—2004 年不同强度蒙古气旋(低压)频数变化

2.3 蒙古气旋(低压)中心的空间分布特征

图 7 为 1948—2004 年蒙古气旋中心位置的空间分布，从图中可以看到，在 105°E 以西，蒙古气旋出现的次数最多，另外，在 46°N，107.5°E 的东南，也是一个蒙古气旋的多发区，而其他区域蒙古气旋出现次数较少，出现次数最少的区域在 115°E 以东。以上的位置分布特点与蒙古气旋(低压)的产生和移动有关。由于 105°E 位于阿尔泰—萨彦岭山地背风坡，因此，这一区域出现较多的蒙古气旋低压大多是由于地形影响出现的背风坡气旋或低压。而 46°N 的东南出现的多发区表明蒙古气旋形成后主要沿东南方向移动，影响我国内蒙古及华北等地，而这也是冷空气爆发南下常常导致气旋向东偏南方向移动的结果。而 46°N 以北和 105°E 以东出现次数较少表明气旋生成后向东偏北方向移动的情况不是很多，同时，多数气旋维持的时间不足以使其移动到 105°E 以东的地区。而这一地区由于处于大兴安岭西侧，也使多数气旋在此减弱或消失。

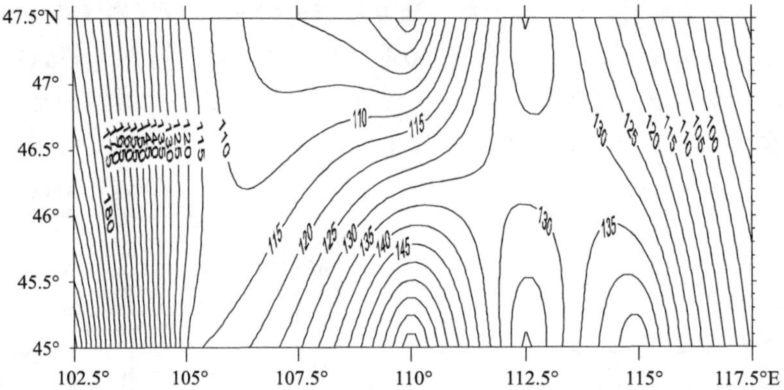

图 7 1948—2004 年蒙古气旋中心位置的空间分布

2.4 蒙古气旋与西风指数的关系

图 8 为 1961—2000 年蒙古气旋发生频数与西风指数距平的演变特征。图中显示，蒙古气旋的发生频数距平与西风环流指数呈现明显的反位相分布；在蒙古气旋次数距平达到极值时，西风指数距平也达到极值，即阶段性的最大值。另外，其变化趋势也是明显呈现反向的；西风

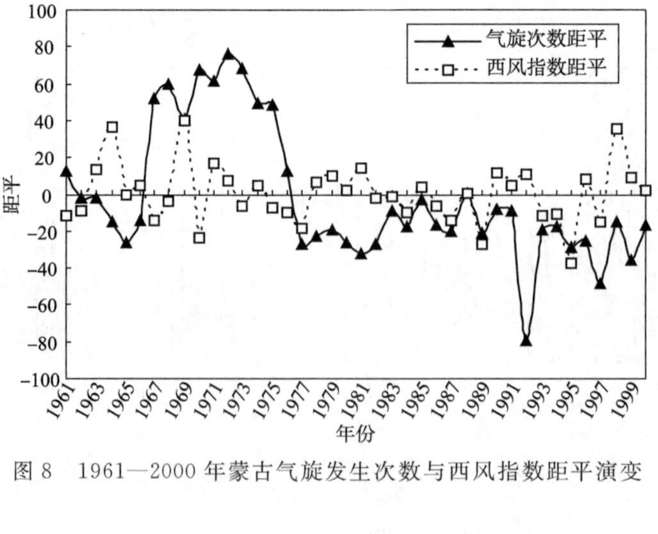

图 8 1961—2000 年蒙古气旋发生次数与西风指数距平演变

指数距平减小时,蒙古气旋发生次数增加。这一结果可以理解为当西风指数距平减小,即中纬度为经向型环流时,蒙古气旋更容易发生。这是因为经向型环流更容易产生冷暖空气的交汇,从而利于气旋产生和增强。

3 结论

利用 NCEP1948—2004 年海平面气压场再分析资料对蒙古气旋(低压)的气候特征进行了统计分析,结果表明:

(1)蒙古气旋(低压)发生频数的高峰出现在 20 世纪 60 年代末 70 年代初,之后呈现减少趋势,目前出现次数处于较低水平。

(2)蒙古气旋(低压)主要出现在春季,然后依次为秋、冬、夏季,并以 5 月为最多。

(3)蒙古气旋(低压)发生频数的年代际特征是 20 世纪 70 年代最多,其次是 60、50、80 年代,90 年代最少。且 60—70 年代出现高峰期主要是由于秋冬季出现次数明显增多所致,春季出现次数并未出现明显的增多。

(4)蒙古气旋(低压)的强度分布基本上接近于正态,强度为 1000～1010 hPa 的出现次数最多,强和特强的(<990 hPa)出现次数相对较少。特别是在近些年出现次数更少。

(5)蒙古气旋(低压)中心主要分布在阿尔泰—萨彦岭山地背风坡,其次为 46°N 的东南。表明气旋生成后主要向东偏南方向移动。

(6)蒙古气旋(低压)频数距平与西风环流指数大致呈现反位相分布,表明当中纬度为经向型环流时,蒙古气旋更容易发生。

参考文献

[1] 王娴. 内蒙古自治区天气预报手册(上). 北京:气象出版社,1987,30-34.
[2] 姜学恭,沈建国,刘景涛,等. 地形影响蒙古气旋发展的观测和模拟研究. 应用气象学报,2004,**5**(15):601-611.
[3] 娜林、姜学恭. 2005 年 4 月一次持续性沙尘暴过程个例分析. 气象,2005,特刊.

Some Climatic Features of Mongolian Cyclones(lows)

JIANG Xuegong[1] KANG Ling[1] HU Yinghua[1] MA Qingxia[2] LI Haiying[2]

1. Inner Mongolia Meteorological observatory, Huhhot 010051;
2. Inner Mongolia Climatic Center, Huhhot 010051

Abstract Using the NCEP sea level pressure reanalysis data from 1948 to 2004, the intermonthly, interannual and interdecadal variation, the strength and occurrence position characteristics of the Mongolia cyclones (lows) are analyzed. It is found that the peak period of the Mongolia cyclones (lows) occurrence is from 1960's to 1970's. The Mongolia cyclones(lows) mostly appear in spring, especially on May. The dominant strength of it is from 1000 hPa to 1010 hPa. The Mongolia cyclones(lows) frequently occur at the lee side of Altai-Sayan complex mountains, next to south of 46° N. The frequency anomaly of the Mongolia cyclones(lows) has a negative relationship to the westerly index. All of these results can be use for the prediction and forecast of Mongolia cyclones.

Key words: Mongolia cyclones(lows), climatic characteristics, statistical analysis

内蒙古两类持续型沙尘暴的天气特征*

姜学恭[1] 沈建国[2]

1. 内蒙古自治区气象台，呼和浩特 010051；内蒙古自治区气象局，呼和浩特 010051

摘　要：利用观测资料对内蒙古地区两种不同类型的持续型沙尘暴过程进行了对比分析，结果表明：天气系统的稳定少动是这两类沙尘暴在同一地区的长时间持续的共同原因。在这 2 次过程沙尘暴的持续期，大气层结的日变化受到不同程度削弱。主要原因是沙尘气溶胶的辐射反馈效应。在冷锋型沙尘暴过程的持续期，地面风速出现的明显日变化加强了沙尘暴强度的日变化。而在高压底部型沙尘暴过程中，暖平流和中尺度低压辐合抬升产生的对流进一步削弱了层结日变化。这一因素以及期间地面风速的相对稳定进一步导致在沙尘暴强度未发生明显改变。

关键词：沙尘暴，天气特征，大气层结

1　引言

近些年，沙尘暴的频繁发生引起了世人的广泛关注，研究者对沙尘天气的发生机理、沙尘天气频繁发生的原因以及沙尘天气的数值预报等进行了深入的研究[1~5]。在沙尘暴的发生发展过程中，层结条件是一个重要的影响因素。王式功等[6]对西北地区强沙尘暴过程的研究发现，在强沙尘暴出现之前往往在大气低层形成不稳定层结，为旺盛干对流的产生创造了有利条件。钱正安等[7]给出的强沙尘暴天气模型把不稳定层结作为一个重要特征。Takemi[8]在研究"9355"黑风暴过程中发现，在黑风暴出现之前，地面加热导致对流层低层形成了深厚的混合层，由于混合层明显减小了抬升气块所需的能量，使得冷锋前地面辐合触发了强干对流和黑风暴。Pauley 等[9]研究影响加州的强沙尘暴过程中也发现在强沙尘暴出现前形成了等熵混合层。可见，地面加热导致大气层结稳定度的降低，并为沙尘暴的发展提供了有利的条件。在强沙尘暴出现之前大气究竟更多地表现为哪种层结形态？则有必要进行进一步的研究。

在沙尘暴发展和维持过程中的层结特点也是一个值得关注的因素。Carlson 等[10]对撒哈拉地区沙尘暴过程的研究发现，在沙尘暴传输过程中，沙尘层（850～500 hPa）维持一等熵混合层。Chen 等[11]通过数值研究发现，沙尘气溶胶的辐射强迫能够在沙尘层中产生一等熵混合层。而在国内，对于沙尘暴维持阶段的大气层结特征还未特别关注。

大风也对沙尘暴的形成和强度具有重要影响。地面大风导致了沙尘的起动。人们对导致沙尘起动的临界地面风速进行了较多的研究。Helgren 等[12]对撒哈拉沙漠西部沙尘暴过程的分析发现，与起沙过程相关联的平均地面风速为 10.5 m/s，而单站的变化范围在 6.5～13.0 m/s 之间。然而，对沙尘暴维持阶段的地面大风特点同样很少研究。本文拟对内蒙古地区 2 种不同类型的持续型沙尘暴过程进行对比分析，通过揭示沙尘暴形成和维持阶段的特点进而揭示沙尘暴形成和维持的一些规律。

* 资助项目　国家自然科学基金资助项目 40565003，科技部社会公益研究专项（2005DIB3J108）资助。

能够引发沙尘暴的天气系统有多种类型[13]。其中,中尺度系统主要为干飑线,这类沙尘暴维持时间较短。在天气尺度系统中,冷锋和气旋是最为常见的,气旋型维持时间相对较长。由于气旋型沙尘暴主要出现在气旋冷锋后部[14],所以冷锋和气旋型沙尘暴具有一定的相似性。另一类维持时间较长的是高压底部偏东风型沙尘暴[14],虽然出现次数较少,但其天气学特点与常见的冷锋和气旋型具有显著差异。因此,本文选取具有普遍意义的气旋型(2002 年 4 月 6—8 日)以及比较特殊的高压底部偏东风型(1994 年 4 月 6—8 日)为例进行持续性沙尘暴特征的对比研究。

2 气旋冷锋型沙尘暴过程

2.1 环流演变及沙尘暴维持原因

2002 年 4 月 6—8 日沙尘暴过程是一次由蒙古气旋强烈发展引发的强沙尘暴过程,沙尘天气出现内蒙古中西部及东部偏南地区、宁夏、山西、河北北部、辽宁、吉林及黑龙江南部,沙尘暴主要出现在内蒙古中部偏北地区(图略)。4 月 5 日 08 时(北京时,下同),蒙古气旋在阿尔泰—萨彦岭山地东侧形成。23 时,气旋东移加强,气旋冷锋侵入内蒙古引发沙尘天气。6 日 08 时后,气旋强烈发展并在内蒙古中部偏北地区导致强沙尘暴。6 日 20 时,气旋进一步加强,中心位于(48°N,120°E),而高空槽快速加深并形成切断,之后气旋移速明显变慢,沙尘暴在内蒙古中部偏北地区持续。7 日 20 时后,随着气旋减弱并快速东移,内蒙古中部偏北地区的沙尘天气也逐渐减弱停止(图 1)。在这次过程中,二连浩特的沙尘天气持续达 2 天之久(6 日 02 时至 8 日 02 时)。

在蒙古气旋型沙尘暴过程中,沙尘天气在同一地点能够维持近 2 天是不多见的。从 4 月 6—7 日平均的海平面气压场(图 1a)可以看到在沙尘天气持续期间气旋中心位于(48°N,120°E)附近,而沙尘天气发生区域(内蒙古中部偏北地区二连浩特附近)始终位于气旋冷锋后部等压线密集区,这一区域往往对应地面大风区,因而具备了持续引发沙尘天气的条件。从环流演变来看,系统发展使高空槽不断加深,在 300 hPa 以下形成了几乎贯穿整个对流层的切断低涡,这样深厚的系统显然其移速变慢。另外,低涡的移动明显呈现非匀速状态。从 5 日 20 时至 6 日 20 时,700 hPa 的切断低涡中心从(50°N,107°E)移至(48°N,119°E),东移 12 个经度;而从 6 日 20 时至 7 日 20 时,低涡中心从(48°N,119°E)移至(47°N,123°E),仅东移 4 个经度;之后从 7 日 20 时至 8 日 20 时,低涡中心从(47°N,123°E)移至(47°N,130°E),东移 7 个经度。在沙尘天气的发生时段(6 日 02 时至 8 日 02 时)系统移动速度是相对较缓慢的。这就导致内蒙古中部偏北地区始终为冷锋后西北气流控制。而系统移动缓慢与高空环流特点有着密切联系。在 6 日 14 时形成切断低涡不久,在 500 hPa 到 250 hPa 间低涡下游的高压形成类似与阻塞形势的“阻塞”高压(图 1b,由于持续时间较短,所以不是严格意义上的阻塞形势),这一“阻塞”形势持续至 7 日 02 时,之后逐渐减弱崩溃,系统移动又逐渐加快。可见,沙尘天气在内蒙古中部偏北地区持续的原因主要是系统旺盛发展,在整个对流层形成深厚的低值系统,导致移动变缓;同时,在对流层中上层形成了类似于“阻塞”的上下游配置,进一步使系统移动变慢。由于内蒙古中部偏北地区能够提供丰富的沙源,在长时间受气旋冷锋后部强的西北气流控制的情况下,产生持续的起沙从而使沙尘天气持续。

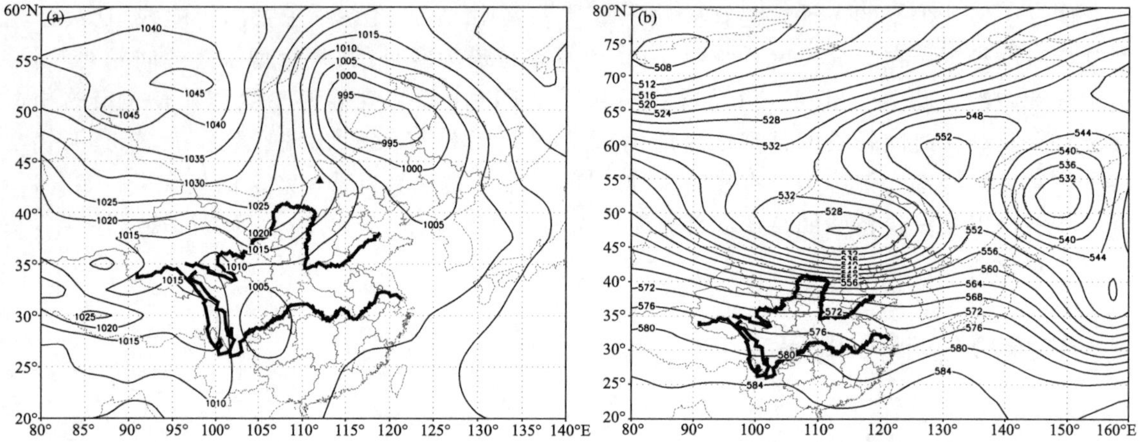

图 1 2002 年 4 月 6—7 日平均的(a)海平面气压场(单位:hPa),(b)500 hPa 位势高度场(单位:dagpm)。其中实三角代表二连浩特位置

2.2 气旋冷锋与地面大风

图 2 为 2002 年 4 月 5—8 日内蒙古中部二连浩特地面气象要素演变。可以看到在 5 日 20—23 时冷锋过境(出现气压上升、温度下降、风速增大、风向由西南转为西北)。之后至 6 日 08 时,上述气象要素的变化最为剧烈,表明为主体冷锋经过。而 6 日 08 之后沙尘天气 2 天多的持续期特征表现为(仅 7 日 02—05 时能见度大于 10 km):温度和气压维持,风向为一致的西北风。从能见度的变化可以看到,沙尘天气强度在持续期出现明显的波动,与其他地面气象要素对比,地面风速的变化与能见度的变化几乎是同步的:呈现严格的反位相关系,当风速增大时,能见度减小,反之能见度增大。在沙尘天气这一持续期,由于天气系统的影响是相对稳定的,因此,在这种情况下,沙尘天气强度与地面风速的关系变得更为单纯了。

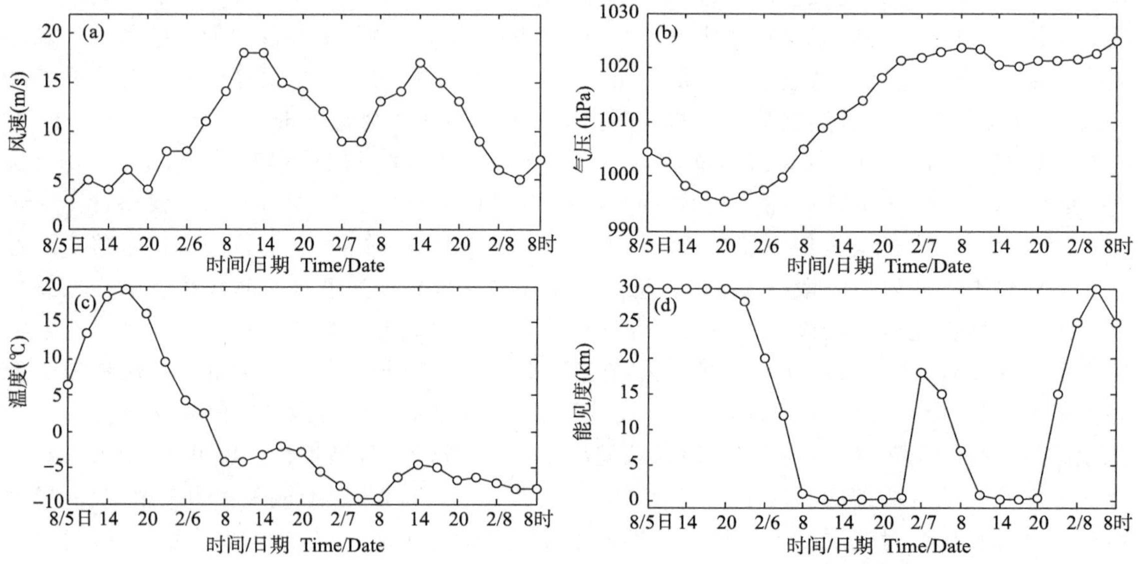

图 2 2002 年 4 月 5—8 日二连浩特地面气象要素演变:(a)风速;(b)气压;(c)温度;(d)水平能见度

另外，在沙尘天气持续期能见度和地面风速的波动也可以视定为日变化：在白天地面风速增大，能见度降低，沙尘暴增强；夜间地面风速减小，能见度增大，沙尘暴减弱。这一现象揭示了沙尘暴强度演变的一些深层原因，首先，地面加热日变化导致的层结日变化对沙尘暴强度具有直接的影响；此外，层结日变化表现为日间混合层，而混合层有利于高空动量下传乃至地面风速增大[14]，因而迫使沙尘暴表现出明显的日变化。

2.3　层结特征

图 3 为 5 日 08 时至 8 日 08 时二连浩特位温和比湿廓线。图中可以看出：首先，在沙尘暴发生前、发展及维持阶段均未出现严格的不稳定层结。其次，在沙尘暴出现前(5 日 08—20 时)，早晨为逆温稳定层，傍晚为地面加热形成的混合层，表现出明显层结日变化，这是常态大气的普遍规律。而沙尘暴出现之后的层结特征明显有别于这一规律。在 6 日 08 时至 7 日 20 时沙尘暴加强和持续期间，混合层始终维持，即在 6 日 08 时和 7 日 08 时本应出现逆温稳定层的早晨也出现了混合层，这一事实表明混合层与沙尘暴之间有着一定的伴随关系(尽管在整个

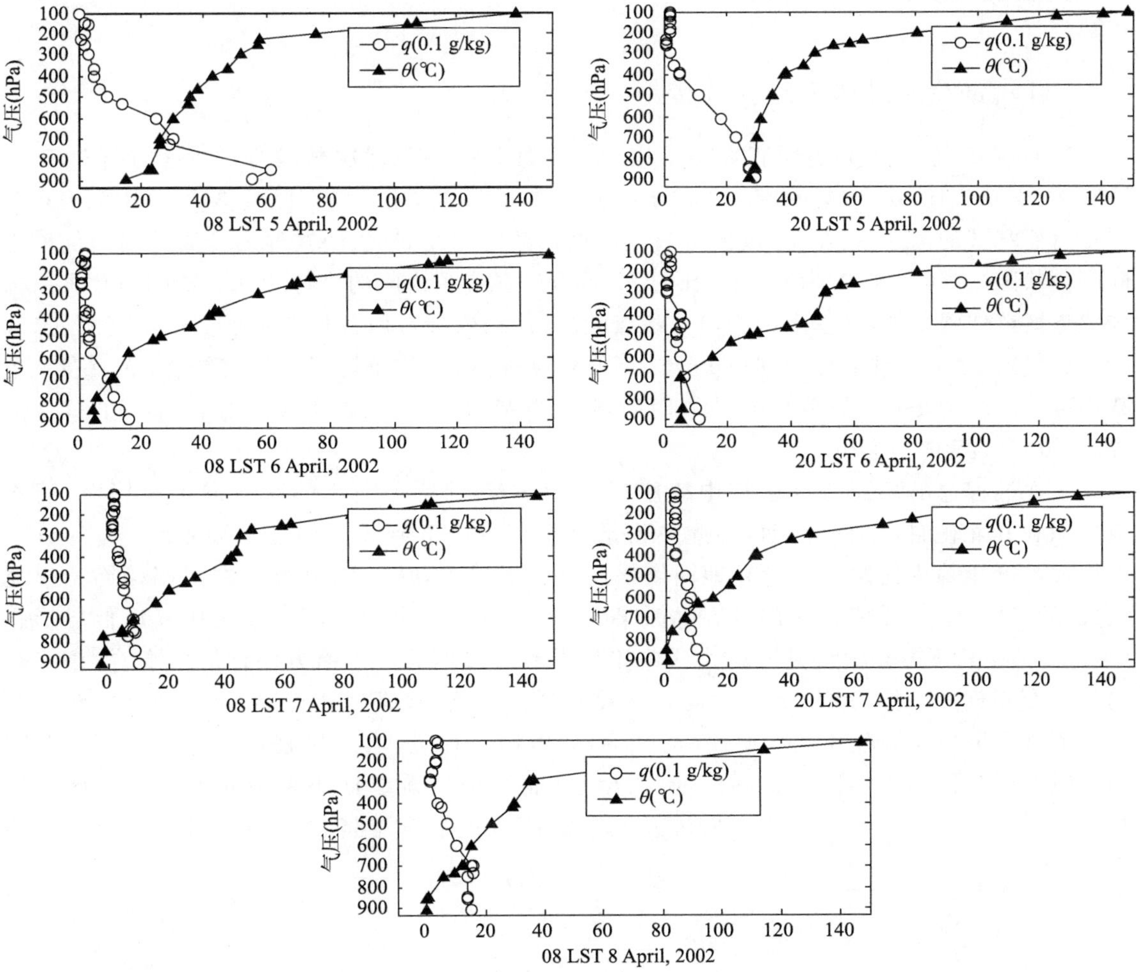

图 3　2002 年 4 月 5 日—8 日二连浩特位温和湿度廓线　位温(θ)，单位：℃，比湿(q)，单位：0.1 g/kg

持续期，混合层表现出一定的高度日变化：在早晨，混合层高度在 800 hPa 左右，到傍晚，混合层抬高至 700 hPa 左右）。同时，沙尘暴强度与混合层的高度表现出相同的变化趋势：在傍晚混合层高度较高时，沙尘暴强度较强，反之强度则较弱。

在沙尘暴发生前，混合层无疑是有利于沙尘暴发生的层结条件[8,14]，但这种关系导致的结果是混合层将先于沙尘暴减弱。在沙尘暴持续期间出现的混合层与沙尘暴密切伴随的现象（7 日 20 时—8 日 08 时），则表明在沙尘暴形成的同时，产生了一种使混合层维持的机制。而这一机制是存在的，那就是沙尘气溶胶的辐射强迫。Carson 等[15]研究表明，在白天，对于海洋或沙漠下垫面，沙尘气溶胶都将减小向下到达地面的辐射通量，同时增加对大气的加热，从而对大气层结施加一种稳定化的影响（stabilizing effect）。同样的道理，在夜间沙尘气溶胶将减小地面向上释放的辐射通量（同时将增加大气向上的辐射冷却），导致的结果是减小夜间逆温稳定的程度。综合来看，沙尘的辐射强迫将削弱大气层结日变化。对于类似本次过程这样的沙尘暴过程而言，这一机制显然有利于混合层的维持。

3 高压底部型

3.1 环流演变及沙尘暴维持原因

1994 年 4 月 6—8 日高压底部偏东风型沙尘暴是一种出现次数较少的沙尘暴天气型[13]（图 4）。4 月 5 日 20 时，蒙古冷高压位于（80°E，54°N）附近，之后向东缓慢移动，其南侧为东西向平直而宽阔的锋区，而在 40°N 附近有暖倒槽自南向北发展，从而在地面冷高压与暖倒槽之间的带状区域（40°—45°N，90°—115°E）形成具有较强气压梯度的准静止锋。期间内蒙古西部偏北地区出现强沙尘暴（本次过程出现站点较少，仅在内蒙古西部偏北地区 2～3 个测站出现沙尘天气）。之后直到 8 日 20 时，由于冷高压移动缓慢，准静止锋得以稳定维持，导致在一个狭小区域内沙尘暴一直持续。8 日 20 时后，暖空气势力减弱，冷高压加强南下，准静止锋崩溃导致沙尘暴逐渐减弱停止。

本次过程总的环流特点是在中纬度（50°—65°N）维持一宽广的东西向平直锋区，而冷空气只在其北侧自西向东移动，因此使得准静止锋区在 6 到 8 日间能够稳定维持（图 4a）。在地面准静止锋南侧暖平流导致地面倒槽向北旺盛发展，从而使得内蒙古西部始终位于地面倒槽顶部。这一低值系统在对流层低层（700 hPa 以下）多数时间表现为中尺度低压形态（图 4b）。在其下游东北侧是同样由于暖平流导致的沿西北—东南方向伸展的高脊，该高脊的位置在 6 日 08 时—7 日 08 时之间几乎没有任何变化，只是到 8 日 08 时才转竖并略有东移。而沙尘天气主要发生在 6 日 14 时至 8 日 08 时。可见，本次过程沙尘暴在内蒙古西部的持续主要是因为中纬度平直的纬向型环流使得冷空气难以南下形成对准静止锋区的破坏，同时，暖平流导致地面倒槽的发展及其下游高脊的稳定维持也为沙尘暴发生提供了动力条件。

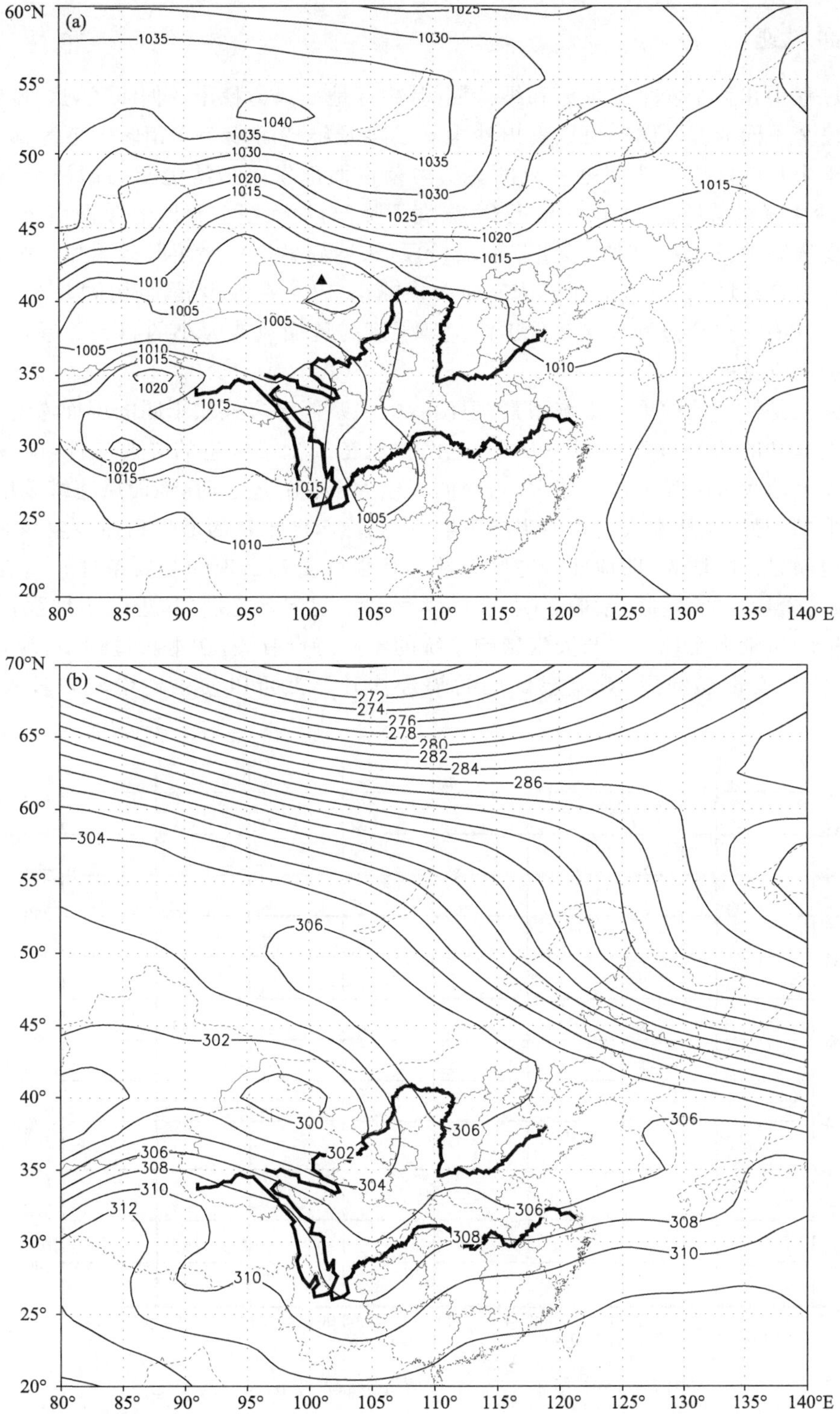

图 4 1994 年 4 月 6—8 日平均的(a)海平面气压场(单位:hPa),(b)700 hPa 位势高度场(单位:dagpm)。其中实三角代表额济纳旗

3.2　地面风速

本次过程中在整个沙尘暴形成和维持期间，主导地面风为稳定的偏东风，这是与常见的沙尘暴天气完全不同的。从图5可以看出沙尘暴发生及维持阶段700 hPa以下为偏东风，与沙尘暴区处于中尺度低压的北部是一致的。从内蒙古西部发生强沙尘暴的额济纳旗可以看出（图6），在沙尘暴维持阶段，出现气压上升、温度下降，与地面图上锋区的缓慢南压对应。而地面风速演变表现为：7日08时前沙尘暴发展阶段（未达到最强），地面风速明显增强（13 m/s）；之后在沙尘暴持续增强阶段，地面风速反而大幅减小，以至在沙尘暴维持阶段，地面风速仅维持在6 m/s左右。整个过程并未表现出沙尘暴强度与地面风速大小的相关关系，这是一般沙尘暴过程中不常见的。

地面大风是沙尘暴的重要影响因素，但两者的关系似乎应该从不同的层面考虑。首先，地面风是沙尘起动的动力因素，在这一层面，地面大风是沙尘暴发生的必要条件，即沙尘暴的发生需要地面风速大于临界风速。其次，当地面风速大于临界风速时，地面风速继续增大固然可能导致沙尘暴增强；但从本次过程的观测事实可以看到，沙尘暴的增强和强沙尘暴的出现并不一定需要地面风速的增大，即地面风速增大并不是强沙尘暴出现的必要条件。这也是容易理解的，因为沙尘暴受到多种因素的影响，在沙尘顺利实现起动之后，其进一步增强可以通过其他影响因素的贡献来实现。这与天气影响系统的不同特点有关，在本次过程中，沙尘在6 m/s风速条件下[12]实现起动后，对流层低层的辐合抬升显然可以使强沙尘暴在较长的时间内维持。

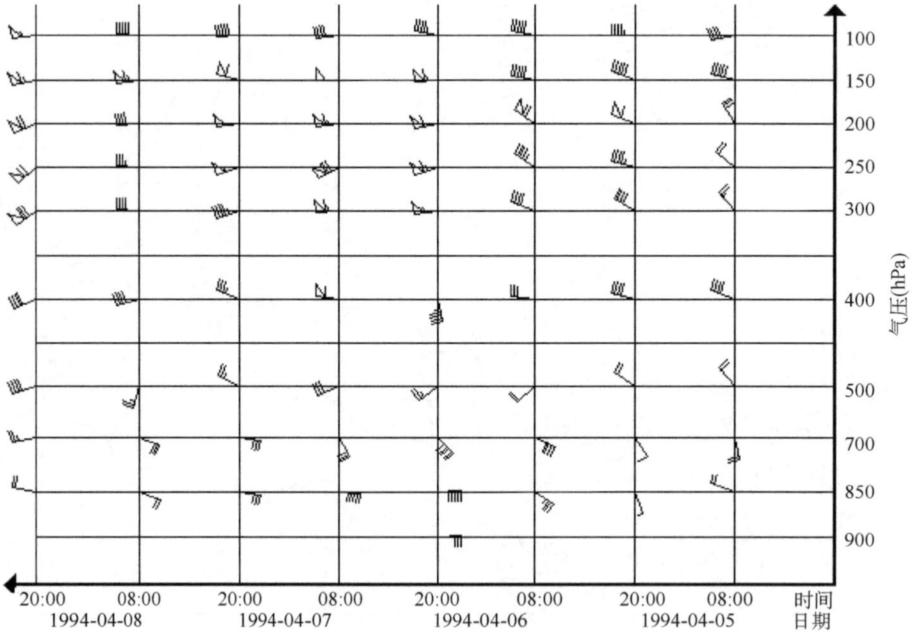

图5　1994年4月5—8日额济纳旗高空测风时间变化

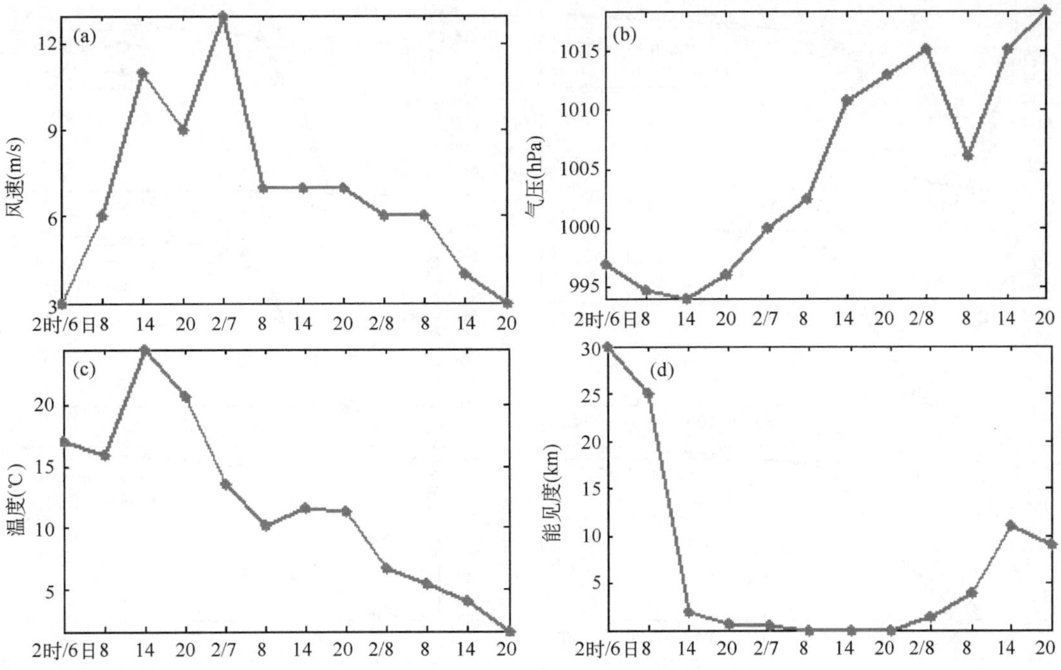

图 6 1994 年 4 月 6—8 日额济纳旗地面气象要素演变

(a)风速;(b)气压;(c)温度;(d)水平能见度

3.3 层结特征

图 7 为 1994 年 4 月 5—8 日额济纳旗位温和比湿廓线。与前一过程类似,在沙尘暴出现前及发展、维持阶段,同样未出现严格的不稳定层结。同时,在 6 日 20 时至 8 日 08 时,沙尘暴持续的同时,混合层也是一直维持的。但这 2 次过程还是存在三点明显的不同:首先,本次过程混合层高度较低(图中显示,从地面到 850 hPa 位温和比湿几乎为常数,显示混合的程度是非常强的),在 850 hPa 左右,而前一过程为 700 hPa;其次,在沙尘暴持续期间,本次过程层结日变化明显弱于前一过程。另外,从图 6 还可看出这一阶段沙尘暴强度未发生明显的日变化,而前一过程沙尘暴日变化是明显的。

本次过程中混合层高度低于前一过程,主要原因是本次过程的影响系统比较浅薄,仅仅存在于 700 hPa 以下,此外也不能排除地形与混合层平流的共同影响[16~18]:当气流方向为下坡方向,则下游地区的混合层高度将被抬高,与上节沙尘扩展方向为下坡过程不同,本次过程低层气流基本为上坡过程。而本次过程层结日变化小于前一过程,是因为本次过程导致混合层维持的因素不仅包括沙尘气溶胶的辐射强迫,同时,地面倒槽的旺盛发展必然与对流层低层的暖平流对应,而对流层低层中尺度低压的存在将导致辐合上升,这两种因素都将有利于垂直对流的产生,进而使大气层结形态表现为混合层形态。因此,本次过程存在两种使混合层维持的机制。而前一过程中只存在前一种使混合层维持的机制,这是因为该过程中沙尘暴区位于气旋冷锋之后,为下沉运动区(诊断分析表明前一过程沙尘暴区为下沉区,而本次过程为上升区,图略),下沉运动显然不能使混合层维持。

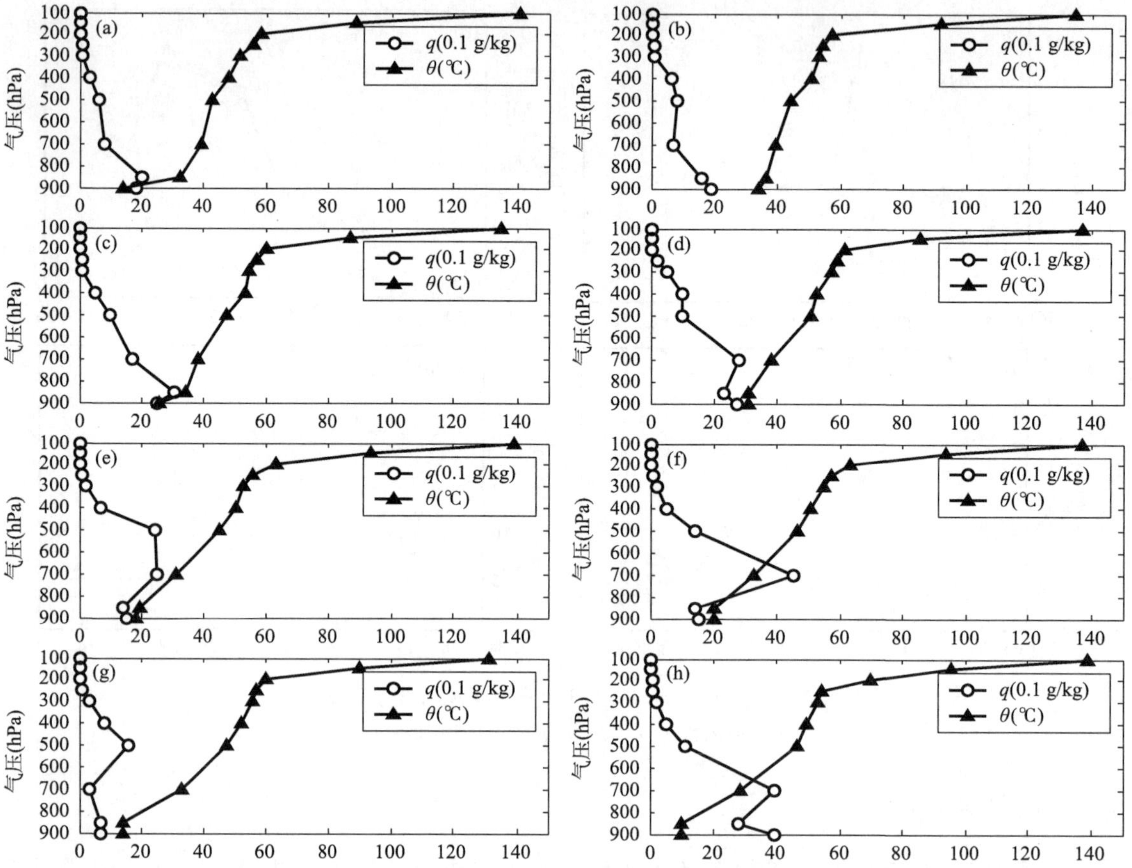

图 7　1994 年 4 月 5—8 日额济纳旗位温和湿度廓线。位温(θ)(单位:℃),比湿(q)(单位:0.1 g/kg)
(a)5 日 08 时;(b)5 日 20 时;(c)6 日 08 时;(d)6 日 20 时;(e)7 日 08 时;(f)7 日 20 时;(g)8 日 08 时;(h)8 日 20 时(地方时)

本次过程沙尘暴强度未发生明显日变化的原因是导致影响沙尘暴的因素都没有日变化,这里最主要是暖平流及中尺度低压产生的辐合抬升是相对稳定的,另外,对沙尘暴具有重要影响的地面风速保持稳定,其大小既大于临界风速,而又未出现波动,因此,不会对沙尘暴强度造成影响。另外,影响沙尘暴的层结因素也未出现日变化。这些因素共同作用使沙尘暴强度未出现日变化。而前一过程影响沙尘暴强度的因素不包含辐合抬升,尽管层结因素的日变化受到部分削弱,但对沙尘暴强度具有重要影响的地面风速却呈现明显的日变化,从而导致沙尘暴强度出现与其几乎同步的日变化。这一现象也说明在影响沙尘暴强度的其他因素保持相对稳定的情况下,地面风速与沙尘暴强度的同步变化突显出了它们之间相对纯粹的相关关系。

4　结论

本文利用观测资料对内蒙古地区两种不同类型沙尘暴过程进行了对比分析,结果表明:

(1)在蒙古气旋型沙尘暴过程中蒙古气旋表现出不均匀的移速,在其缓慢移动期间沙源区始终位于气旋冷锋后部强西北气流控制的区域,从而导致沙尘暴在这一地区持续。而导致其移速缓慢的原因是系统的旺盛发展以及形成了类似“阻塞”的上下游配置。

(2)在高压底部型沙尘暴过程中,平直的纬向型环流为准静止锋区和沙尘暴的维持提供了适宜的条件；同时，暖平流导致地面倒槽和对流层低层中尺度低压旺盛发展为沙尘暴发生提供了动力条件。

(3)在两类沙尘暴过程的持续期，常态大气条件下的层结日变化受到不同程度的削弱。它们共同的影响因素是沙尘气溶胶的辐射强迫效应。这一强迫将同时削弱白天的层结不稳定度和夜间的层结稳定度。

(4)在冷锋型沙尘暴过程的持续期，层结日变化被削弱的程度较小，同时，地面风速出现很强的日变化，从而导致沙尘暴呈现与地面风速一致的强度日变化。

(5)在高压底部型沙尘暴过程中，层结日变化基本被消除。这是因为暖平流和中尺度低压产生的辐合上升和对流对大气施加了另外一种使混合层维持的机制。

(6)在高压底部型沙尘暴的持续期，影响沙尘暴强度的一些因素，包括地面风速、辐合抬升及层结均未发生明显变化，因而沙尘暴强度也未发生明显变化。

参考文献

[1] 吴晓京，郑新江，李小龙等. 东亚春季沙尘天气的卫星云图特征分析和分型. 气候与环境研究，2004，**9**(1)：1-13.

[2] 孙建华，赵琳娜，赵思雄. 一个适用于我国北方的沙尘暴天气数值预测系统及其应用试验. 气候与环境研究，2003，**8**(2)：125-142.

[3] 邵亚平. 沙尘天气的数值预报. 气候与环境研究，2004，**9**(1)：127-138.

[4] 郑新江，扬义文，李云. 北京地区沙尘天气的某些特征分析. 气候与环境研究，2004，**9**(1)：14-23.

[5] 刘景涛，郑新江，康玲，等. 蒙古气旋爆发性发展导致的强沙尘暴个例研究. 气候与环境研究，2003，**8**(2)：218-229.

[6] 王式功，杨德保. 我国西北地区黑风暴的成因和对策. 中国沙漠，1995，**15**：19-30.

[7] 钱正安，蔡英，刘景涛. 中国北方沙尘暴研究若干问题. 沙尘暴学术研讨会文集，台湾：台湾蒙藏委员会出版，2002. 63-76.

[8] Takemi T. Structure and evolution of a severe squall line over the arid region in northwest China. *Mon. Wea. Rev.*，1999，**127**：1301-1309.

[9] Pauley P M，Baker N L，Barker E H. An observational study of the"interstate 5"dust storm case. *Bull. Amer. Meteor. Soc.*，1996，**77**：693-719.

[10] Carlson T N，Prospero J M. The large-scale movement of Sahalan air outbreak over the northern equatorial Atlantic. *J. Appl. Meteo.* 1972. **11**：283-297.

[11] Chen S J，Kuo Y H. The effect of dust radiative heating on low-level frontogenesis. *J. Atmos. Sci.*，1995，**52**：1414-1420.

[12] Helgren D M，Prospero J M. Wind velocities associated with dust deflation events in the western Sahara. *J. App. Meteo.*，1987，**26**：1147-1151.

[13] 刘景涛，钱正安，姜学恭等. 中国北方特强沙尘暴的天气系统类型分析. 高原气象，2004，**23**(4)：540-548.

[14] 姜学恭，沈建国，刘景涛，等. 导致一例强沙尘暴的若干天气因素的观测和模拟研究. 气象学报，2003，**61**(5)：606-620.

[15] Carlson T N，Benjamin S G. Radiative heating rates for Saharan dust. *J. Atmos. Sci.*，1980，**37**：193-213.

[16] Stull R B. 边界层气象学导论 见：徐静琦，扬殿荣 译. 青岛：青岛海洋大学出版社，1991. 1-457.

[17] Arritt R W，Wilczak J M，Young G M. Observations and numerical modeling of an elevated mixed layer.

Mon. Wea. Rev.,1992,**120**:2869-2880.

[18] Carlson T N,Benjamin S G,Forbes G S. Elevated mixed layer in the regional severe storm environment: Conceptual model and case studies. *Mon. Wea. Rev.*,1983,**111**:1453-1473.

The Characteristic of Weather Factors in Two Types of Sustained Dust Storm in Inner Mongolia

JIANG Xuegong[1]　Shen Jianguo[2]

1. Inner Mongolia Meteorological Observatory, Huhhot 010051;

2. Inner Mongolia Meteorological Bureau, Huhhot 010051

Abstract By using the observational data, a comparative study on the weather characteristic within the maintaining period of two different types of dust storm was conducted. The result show that in these two cases the dust storms sustained for two or three days. It is due to the weather influence systems stood or moved slowly. The diurnal change of stratification was minimized in two different degrees. The common reason was the radiation of dust aerosol. In the Cyclone typical case, the intensity of dust storm exhibited an obvious diurnal change owing to the diurnal change of the strength of surface wind speed. In High Bottom case (the dust storm appeared at the bottom of the surface high), the meso-low (under 700 hPa) conducted a key influence on the dust storm through its induced ascending motion. It also contributed to the sustaining of mixed layer. Meanwhile, the surface wind speed also did not changed in the sustaining period of dust storm. And so in the High Bottom case the dust storm maintained with same intensity for more than two days.

Key words: dust storm, weather characteristic, atmospheric stratification

三类沙尘暴过程环流特征及动力结构对比分析*

姜学恭　云静波

内蒙古自治区气象台，呼和浩特　010051

摘　要：冷锋、蒙古气旋、高压底部倒槽型是导致东亚和我国北方沙尘暴的3类主要天气尺度系统。为了揭示3类过程的异同并为沙尘暴预报提供参考，本文利用天气观测资料对3类过程的大尺度环流、系统动力热力结构特征进行了对比分析，结果表明：从大尺度环流特征来看，冷锋型、蒙古气旋型均在高纬地区形成尺度较大的槽涡，槽涡底部出现低槽分离并向南发展，在中纬度诱发地面冷锋及蒙古气旋。其差异在于蒙古气旋过程中往往在中纬度对流层中低层形成切断低涡。蒙古冷高压底部倒槽型过程中，中高纬为脊前西北气流，中纬度蒙古冷高压与向北发展的倒槽在中纬度遭遇形成准静止锋，并在其南侧诱发沙尘暴。从动力、热力结构来看，冷锋型、蒙古气旋型在对流层中低层均呈现典型的斜压结构，冷锋型过程锋区异常陡立，700 hPa以下近于垂直；上升运动呈倾斜形态，并在对流层中低层形成高值中心。蒙古气旋过程中，气旋区形成8—10个纬距上升气柱，贯穿整个对流层。蒙古冷高压底部倒槽型过程中，沿经向700 hPa以上则形成南北风的明显交汇，而在其下方形成南侧沙尘区上升、北侧高压区下沉的垂直正环流。通过进一步对比三类过程的发展和演变规律给出了预报着眼点。

关键词：沙尘暴，天气系统，大尺度环流，动力热力结构，对比分析

1　引言

沙尘暴是发生在干旱、半干旱荒漠化地区的重要灾害性天气，沙尘暴发生后，通过远距离传输对下游地区工农业生产和人民生活造成巨大的影响，并直接影响空气质量、生态环境乃至气候变化。目前，沙尘暴已不仅仅作为气象灾害受到关注，而是成为严重的环境问题。

沙尘暴的发生需要地面强风、适宜的层结条件和沙尘源等基本条件。岳平等对民勤一次沙尘暴过程的分析揭示了沙尘暴发生前后的层结演变特征[1]。陈楠等对宁夏不同强度沙尘暴环流特征的对比分析表明，一脊一槽和西北路冷空气是宁夏强和特强沙尘暴的主要环流类型，而天气系统及其强度差异是沙尘暴强度差异的形成原因[2]。王可丽等对2001年我国北方5次沙尘暴环流动力结构的分析表明，蒙古国低值系统、乌拉尔山高压及其之间的气压梯度是沙尘暴的动力源[3]。郑新江等对我国西北的一次沙尘暴过程分析指出，在大气环流调整期间，斜压槽和冷锋的发展触发了沙尘暴[4]。

以上这些研究从环流、动力结构等不同角度分析了沙尘暴的成因。而对沙尘暴形成具有直接和重要影响的是天气系统。Brazel等对美国亚利桑那州沙尘暴过程的历史统计分析

* 资助项目：国家自然科学基金项目（40965005）和内蒙古自然科学基金项目（2010MS0613）。

表明，锋面系统、雷暴与对流、热带扰动和上层切断低压等均能够导致沙尘暴[5]。王式功研究指出，适宜沙尘暴发生的环流型和天气系统包括经纬向环流调整时期、冷锋活动、低空东风急流、中尺度系统等[6]。Pauley 等则认为，与地形密切联系的背风坡槽、陡峭地形附近形成的地形强迫地面风，均适宜沙尘暴的形成[7]。这些研究中涉及的系统既有天气尺度，也有中尺度，而着眼点既有大尺度环流，也有次天气尺度急流等。刘景涛等着眼于天气影响系统，利用 1957—1996 年 40 年高空和地面天气图资料较全面地分析了导致我国北方沙尘暴的天气系统类型，指出，冷锋、蒙古气旋、蒙古冷高压底部倒槽、干飑线是导致我国北方沙尘暴的主要天气系统，基本上概括了我国北方沙尘暴过程的总体特征[8]。这四类天气系统中，干飑线是中尺度系统，且出现次数较少；冷锋、蒙古气旋是天气尺度系统，蒙古冷高压底部倒槽是准天气尺度系统，这 3 类系统具有相似的空间尺度，且发生频数大(冷锋和蒙古气旋)、持续时间长、影响地区广。近年来，尽管研究者对我国北方沙尘暴过程开展了较多的天气学研究，但针对这 3 类过程动力、热力结构的对比分析研究还不是很多。因此，本文拟选取这三类影响较大的天气尺度(准)影响系统，利用典型个例对其动力、热力结构特征进行对比分析，揭示它们导致不同沙尘暴天气特征的内在原因，从而为沙尘暴预报提供参考和借鉴。

2　个例和资料

本文针对冷锋型、蒙古气旋型、蒙古冷高压底部倒槽型三种类型沙尘暴过程进行研究，个例分布选取 2004 年 3 月 9—10 日冷锋型沙尘暴过程、2001 年 4 月 6—8 日蒙古气旋型沙尘暴过程及 1994 年 4 月 6—8 日蒙古冷高压底部倒槽型沙尘暴过程。使用资料为常规高空、地面气象观测资料、1°×1°NCEP 再分析资料等。

发生在东亚及我国北方的沙尘暴过程中，蒙古气旋型和冷锋型具有一定程度的相似特征，首先，两者均出现地面冷锋，或为纯冷锋，或为气旋冷锋；其次，由于蒙西山地(萨彦岭山地)的背风坡效应，两类过程中通常均会出现地面低压。为了准确区分蒙古气旋型和冷锋型，通过研究发现两者存在如下差异：①纯冷锋位于地面高、低压之间的过渡带，未穿过低压中心；气旋冷锋穿过低压中心，并存在显著的气旋暖锋。②纯冷锋过程中，沙尘天气出现在冷锋后，与地面偏北风或偏西风配合，低压区附近没有沙尘天气发生；蒙古气旋过程中，沙尘天气不仅出现在气旋冷锋后，也出现在低压(气旋)区，包括其东南、东北象限，可与地面偏南风或东南风配合。③蒙古气旋过程中，遥感图上沙尘云呈现涡旋带状。而冷锋型则多为斜压叶状。

3　冷锋型沙尘暴过程

3.1　天气过程概括和大尺度环流特征

2004 年 3 月 9 日沙尘暴过程是由冷锋引发的强沙尘暴过程，对我国北方造成巨大影响，其影响范围包括蒙古国，我国西北、华北、东北、甚至长江中下游地区。图 1 为本次过程强沙尘暴发生时的地面天气图，可以看到，冷锋呈现准南北向，伸展达到上千千米，沙尘暴区位于冷锋后部,且与地面大风区配合。本次过程在东亚地区共有77站出现沙尘暴,过程最小能见度100 m,过程最大风速22 m/s在宁夏、内蒙古等地造成了较大的财产损失。

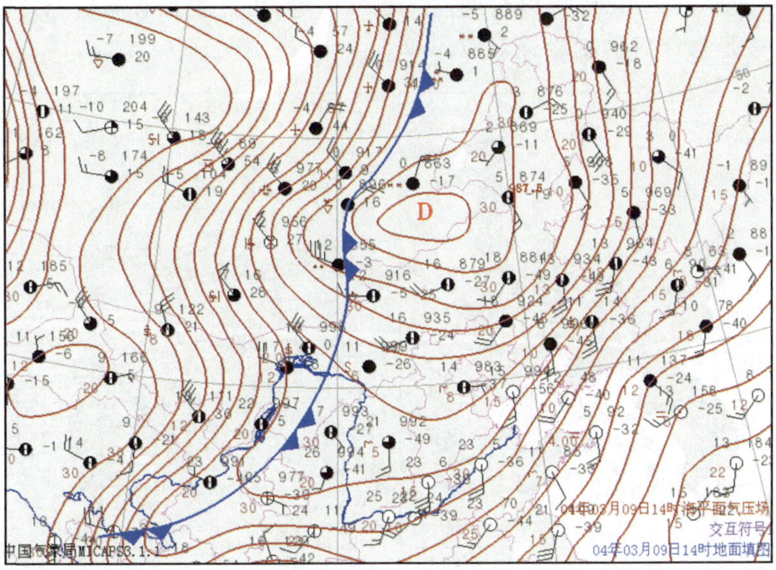

图1 2004年3月9日14时地面天气图

从高空环流形势可以看到(图2),欧亚大陆为两脊一槽型,从乌拉尔山到蒙古东部为宽广的低槽,且在高纬(60°—70°N)表现为尺度较大的低涡,出现在从200 hPa到850 hPa的整个对流层。低涡底部有低槽分离并向南发展,随着高度降低,低槽逐渐加深(在850 hPa形成切断),向南伸展至40°N附近。从温度场来看,200 hPa温度锋区为西南—东北走向,随高度降低逐渐变平,到850 hPa锋区呈东西向,锋区强度在对流层中下层异常强盛,500 hPa在40°—45°N间温度梯度达到20℃;温度槽落后于高度槽,表现为典型的斜压结构,等高线与等温线交角随高度降低而增大,500 hPa为45°左右,而850 hPa接近垂直。可以看出,大尺度低涡底部低槽的分离及其向南扩展,导致了高空锋区的南压,引发强冷空气的爆发。

3.2 动力、热力结构特征

图3为2004年3月9日20时穿过冷锋经向垂直剖面图,可以看到,地面冷锋位于37°N附近,在对流层中,锋区随高度增加向北倾斜,锋区异常陡立,在700 hPa以下接近垂直。暖区内存在两个较强的次级环流(反环流),分别位于925～700 hPa以及300～600 hPa间。从垂直速度分布来看,与锋前次级环流对应,锋前接近锋面区域存在两个明显的上升区,中心分别位于500 hPa和750 hPa,强度分别为−0.7 Pa/s和−0.6 Pa/s,锋后下沉运动中心位于700 hPa附近,强度接近锋前上升运动。考虑到沙尘暴过程中,沙尘粒子主要集中在700 hPa以下,因此,中心位于750 hPa的强上升运动区对于沙尘扬升具有更为重要的意义。散度的垂直分布特征与垂直速度区相对应,其中,与锋前750 hPa的强上升区对应的低层辐合区的中心位于850 hPa,位于辐合中心正上方的辐散中心在600 hPa,与700 hPa以下几乎垂直于地面的锋面是一致的,显示了冷锋的强盛。在对流层中层,辐合和辐散区范围明显扩大,中心区分别

位于 600 hPa 和 400 hPa，辐合区稍落后于辐散区，位于锋区附近，强度也明显强于辐散区。可见，锋前上升运动区呈现明显的梯次分布，因而可以推断沙尘粒子的扬升也明显的分为两个阶段，这与沙尘暴过程中沙尘主体位于行星边界层中，而粒径较小的沙尘往往能够到达对流层中上层的特点是一致的[9]。另外，可以看到，在 43°N 附近，可以看到有与以上冷锋动力结构相似的垂直环流，应该是与锋后的副冷锋对应。

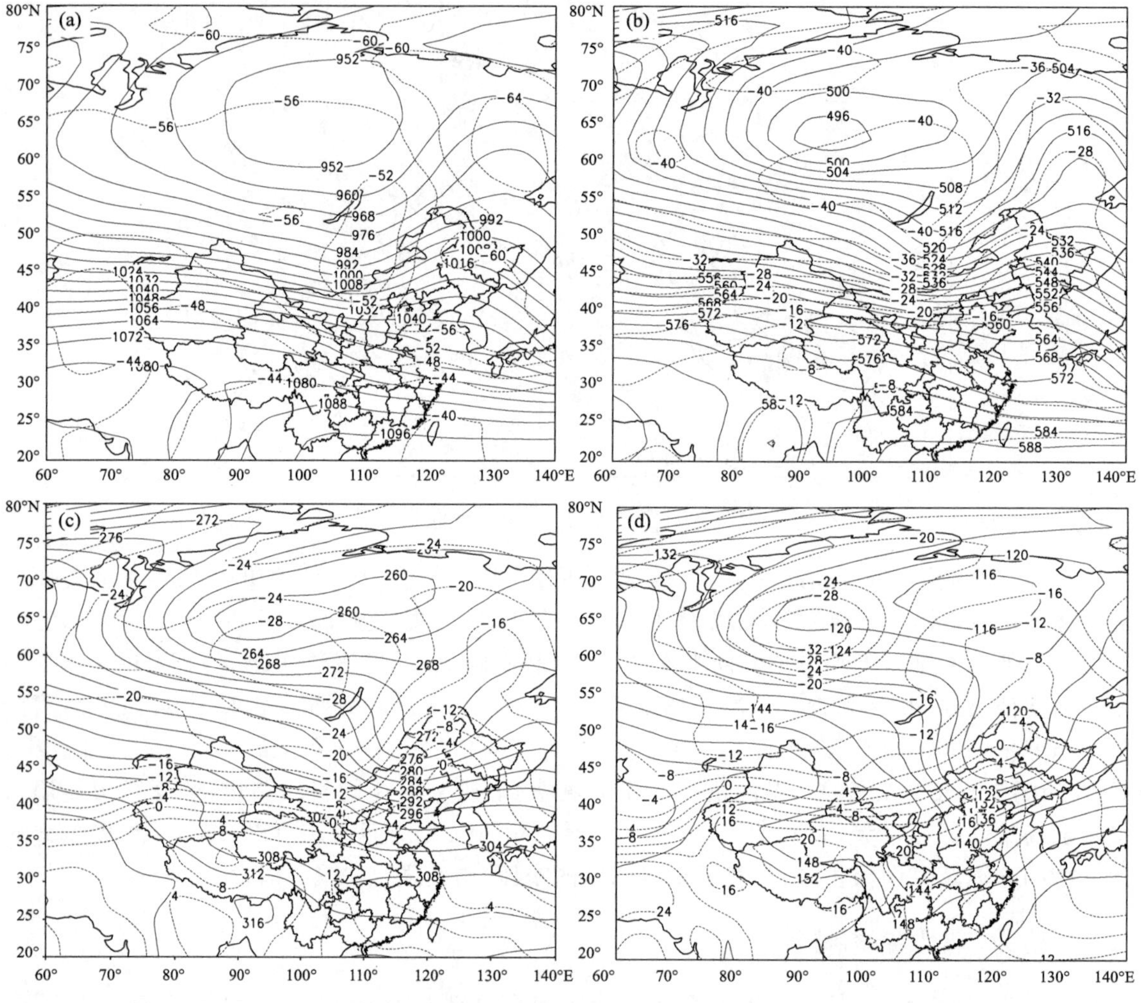

图 2　2004 年 3 月 9 日 20 时高空形势图，其中，a、b、c、d 分别为 250、500、700、850 hPa，实线为位势高度（单位：dagpm），虚线为温度（单位℃）

另外，可以看到，锋前 600 hPa 以下为相对湿度小于 30％的干区，600 hPa 以上湿度条件有所增强，但相对湿度大于 70％的区域很小。而锋后相对湿度更低于锋前。可见，即时在冷气团区域，冷锋型沙尘暴过程中大气的相对湿度仍然是较低的。

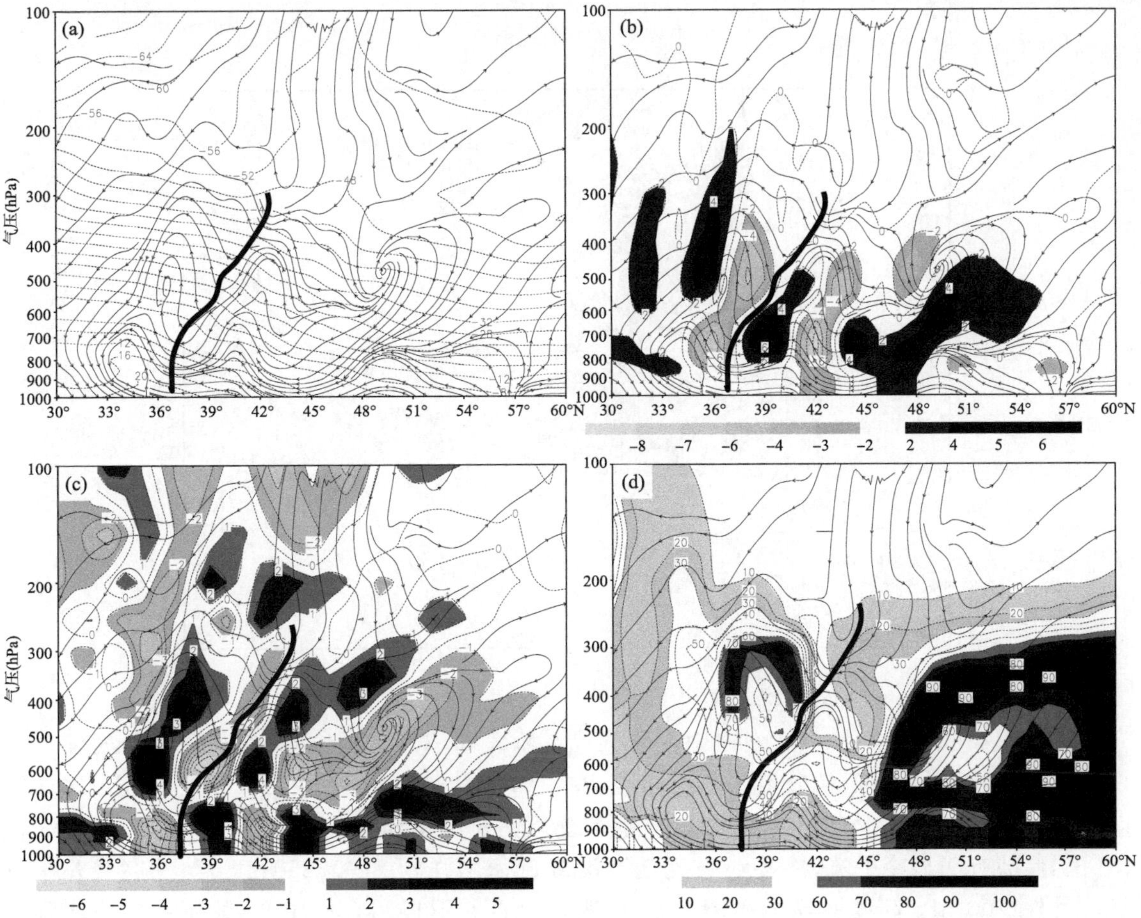

图 3　2004 年 3 月 9 日 20 时过冷锋(110°E)经向垂直剖面图,风矢为 v、w 合成,单位 m/s,10^{-2} Pa/s,等值线分别为(a)温度(K),(b)垂直速度(10^{-1} Pa/s),(c)散度(单位 10^{-5} s^{-1}),(d)相对湿度(%),粗实线为锋区

4　蒙古气旋型沙尘暴过程

4.1　天气过程概括和大尺度环流特征

2001 年 4 月 6—8 日沙尘暴过程是一次影响较大的大范围特强沙尘暴过程,由蒙古气旋引发,波及地区包括蒙古国,我国西部、华北、东北等多个省区,东亚地区共 30 个测站出现沙尘暴,过程最小能见度 0 m,出现在内蒙古中部的 5 个测站,过程最大风速 20 m/s。本次过程仅在内蒙古中部农牧区即造成巨大的经济和财产损失。图 4 为 2001 年 4 月 6 日 17 时地面天气图。

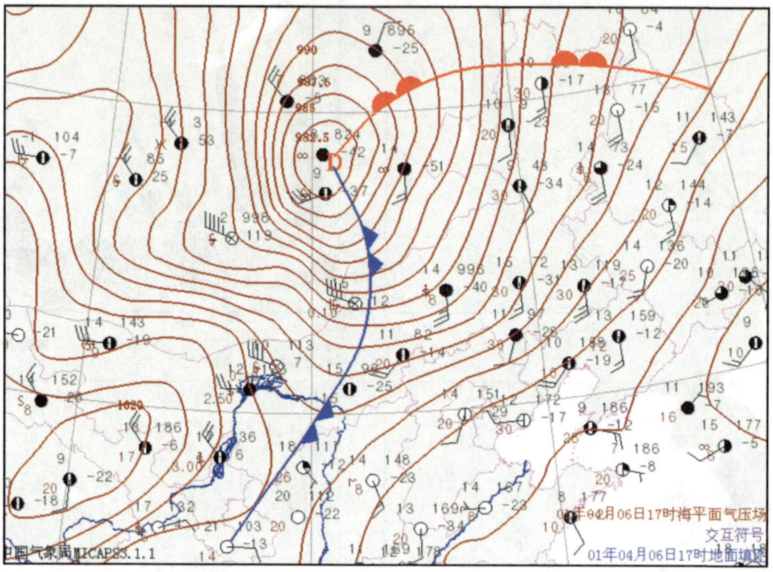

图 4　2001 年 4 月 6 日 17 时地面天气图

图 5 为 2001 年 4 月 6 日 20 时蒙古气旋成熟阶段高空形势图，在高纬地区整个对流层表现为大槽，大槽底部有低槽分离并向南伸展，在中纬度发展加深，并在对流层中低层（500 hPa 以下）形成切断低涡。与温度场的配置显示系统呈明显的斜压结构，温度场落后于高度场，低槽（涡）后部有明显冷平流，冷平流强度随高度降低而增强，在 700、850 hPa 与等高线几乎垂直。另外，200 hPa 高度槽较浅，但槽区出现明显冷舌并形成闭合冷中心，显示高空锋区强盛并有明显南压趋势，这将对气旋发展带来明显的高层动力强迫。

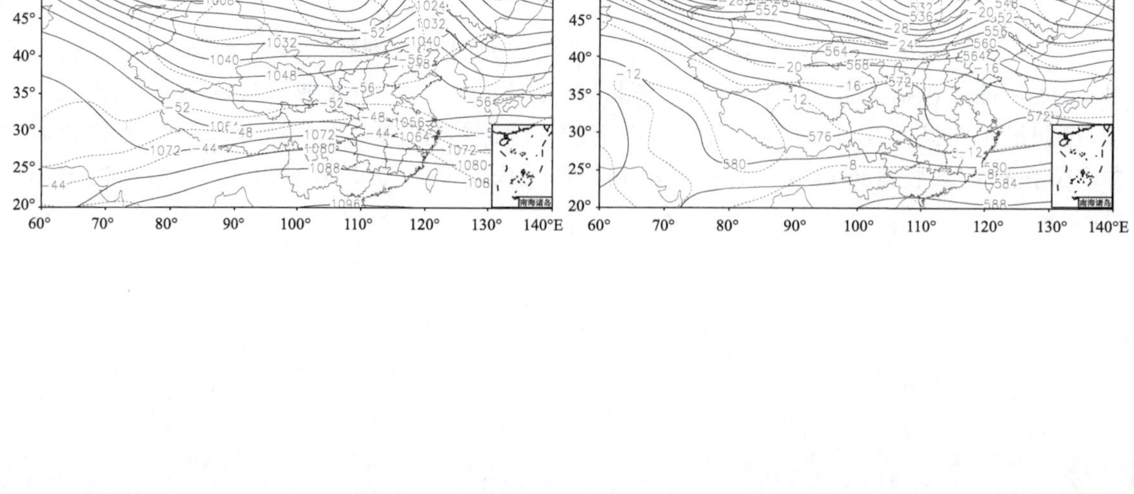

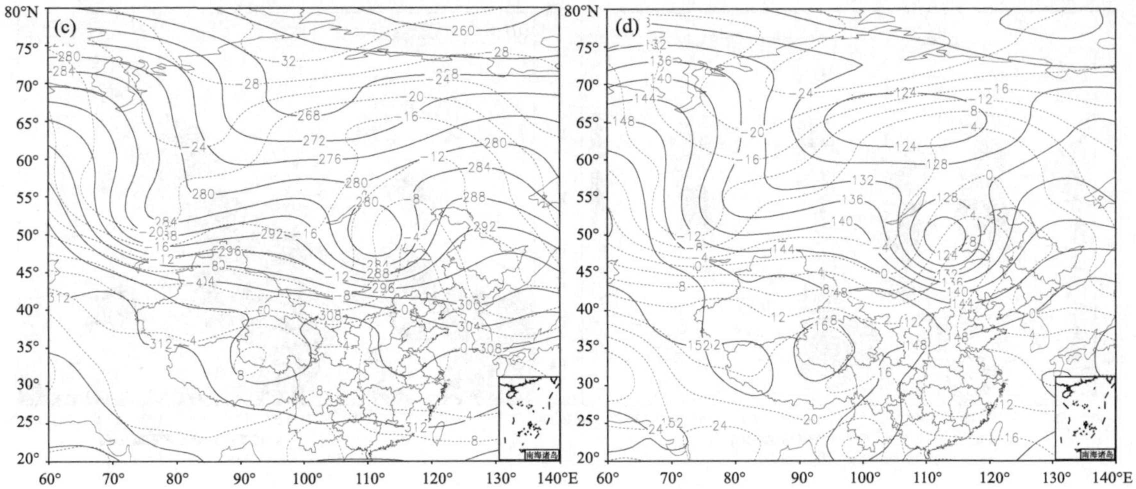

图 5 2001 年 4 月 6 日 20 时高空形势图，实线为位势高度，虚线为温度，其中，a、b、c、d 分布为 250、500、700、850 hPa

4.2 动力、热力结构特征

图 6 为沙尘暴旺盛发展阶段穿过蒙古气旋中心的纬向剖面图(50°N)，首先，可以看到在气旋中心附近从 900 hPa 到 250 hPa 整个对流层为一个巨大的上升气柱，从 105°—115°E 间大约横跨 8～10 个经度，上升运动中心位于气旋中心(112°E)正上方的 650 hPa 附近，强度达到 10 Pa/s 以上。从散度分布来看，低层辐合中心位于气旋中心上方 850 hPa 附近，为偏西气流和偏东气流交汇区，而高层辐散区存在两个中心，一个位于气旋中心正上方的 300～400 hPa 间，呈柱状形态，另一个位于气旋下游(气旋中心以东约 6 个经距)的 250 hPa 附近。可见，该辐散区为气旋区高层气流汇入 250 hPa 高空急流而形成的，图中可以看到该辐散区的存在使得气旋区接近垂直的上升气柱在对流层高层明显变宽，并呈斜升形态汇入高空急流。从涡度分布来看，与上升气柱配合的为正涡度区，并形成贯穿整个对流层的垂直涡柱，其中，正涡度中心

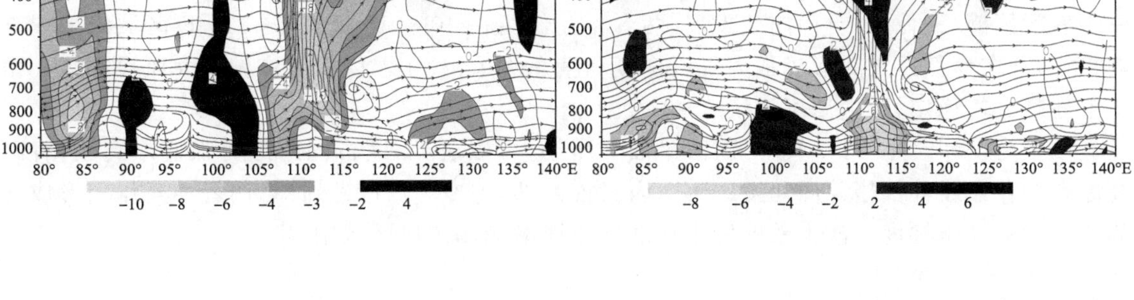

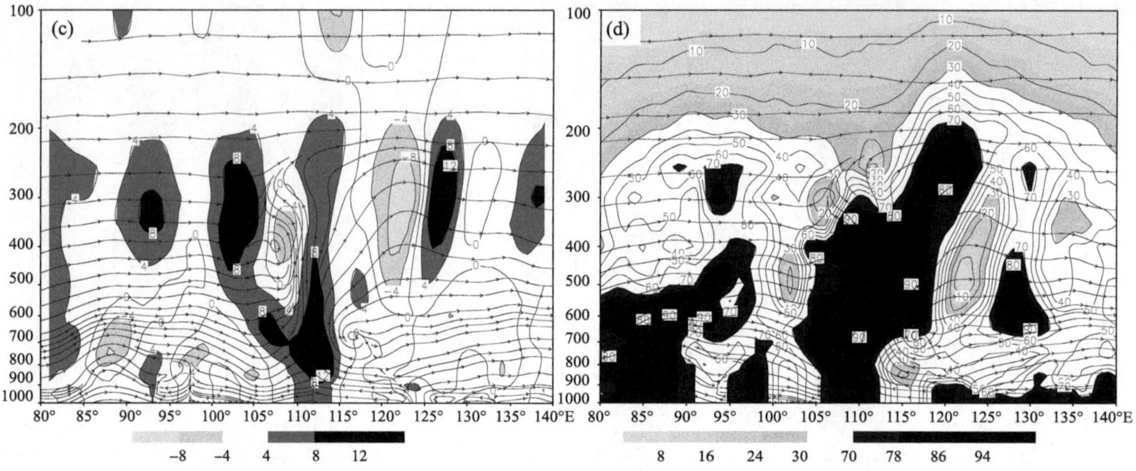

图6　2001年4月6日20时穿过气旋中心(48°N)的纬向垂直剖面图 风矢为u、ω合成,单位m/s,10^{-2} Pa/s,等值线分别为:(a)垂直速度(Pa/s),(b)散度(单位$10^{-5}\ s^{-1}$),(c)涡度,($10^{-5}\ s^{-1}$),(d)相对湿度(%)

位于750 hPa附近。从相对湿度垂直分布来看,在气旋区垂直涡柱的300～700 hPa间存在相对湿度大于90%的区域,但与常见的降水过程的湿度分布特征的明显差异是:在对流层低层700 hPa以下流入气流均为相对湿度小于30%的干区,表明对流层低层大气干燥,而对流层中层相对湿度大值区的形成显然是由于气块上升温度降低所致,其本身的水汽含量并不高,这也是过程难以形成有效降水的原因。

5　蒙古冷高压底部倒槽型沙尘暴过程

5.1　天气过程概括和大尺度环流特征

蒙古冷高压底部倒槽型是一种出现次数较少的沙尘天气过程。与通常由冷锋或气旋冷锋驱动的沙尘暴天气过程不同,这类沙尘暴过程是由准静止锋引发的,1994年4月6—8日内蒙古中西部沙尘暴过程就属于这一类型。这次过程仅在内蒙古西部额济纳旗、拐子胡等3个测站引发沙尘暴,影响范围较小,但沙尘暴持续时间达到3天左右,是一种典型的次天气尺度局地沙尘暴过程[10]。

这类过程天气系统活动特征表现为:蒙古国附近存在强大的冷性高压,其前部冷锋在120°E以西由东北—西南向转为准东西方向,并稳定少动,同时,在我国河西走廊到河套地区一带有暖倒槽(或低压)自南向北发展,两者相互作用在40°—45°N、90°—115°E区域内形成准静止锋,并在高低压之间形成强大气压梯度和偏东大风,沙尘暴就发生在偏东大风控制区。由于上述环流较稳定,沙尘暴可持续2～3天。

图7为1994年4月7日08时海平面气压场,可以看到,蒙古国附近为强盛冷高压,青藏高原东侧存在暖性倒槽并向北发展直达内蒙古西部,与冷高压南北对峙,两者相互作用形成准静止锋和强气压梯度。沙尘暴即发生在准静止锋南侧偏东风区域中。

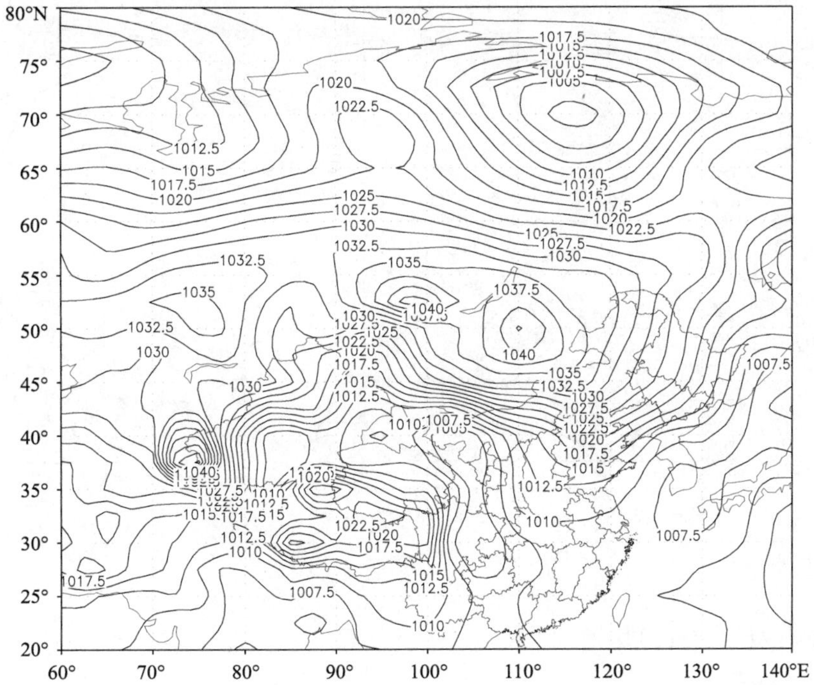

图 7　1994 年 4 月 7 日 08 时海平面气压场

图 8 高空形势显示，欧亚大陆维持两槽一脊，乌拉尔山东侧及亚洲东岸为低槽，之间则为宽广的高压脊，我国北方位于脊前西北气流控制地区，对流层中低层锋区强盛，并随高度降低逐渐从西北—东南走向变为准东西走向，维持在蒙古南部到内蒙古中西部一带，与地面的准静止锋对应。同时，在对流层低层蒙古国附近出现阻塞高压（近似于阻塞形势），这种较为稳定的环流型配置显然有利于准静止锋的维持。

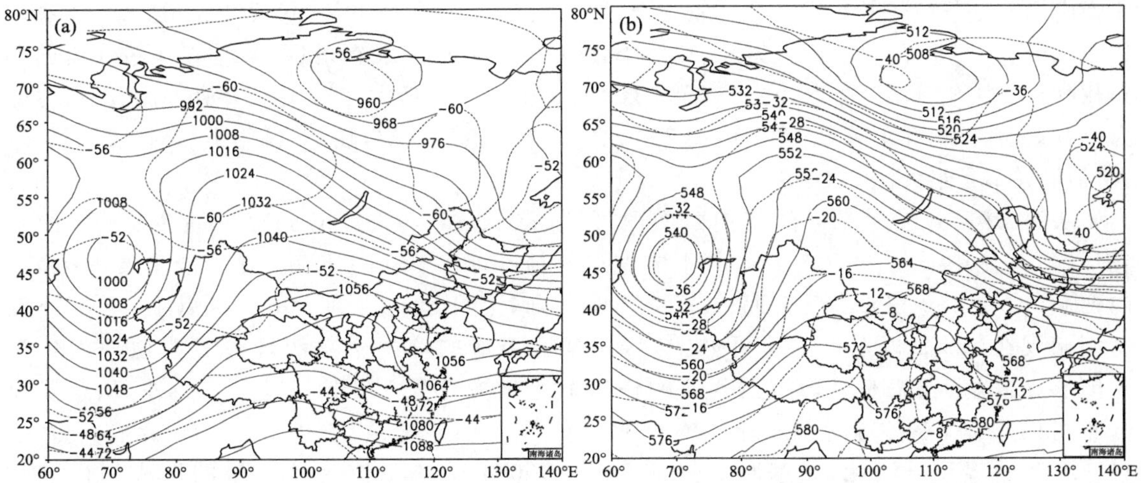

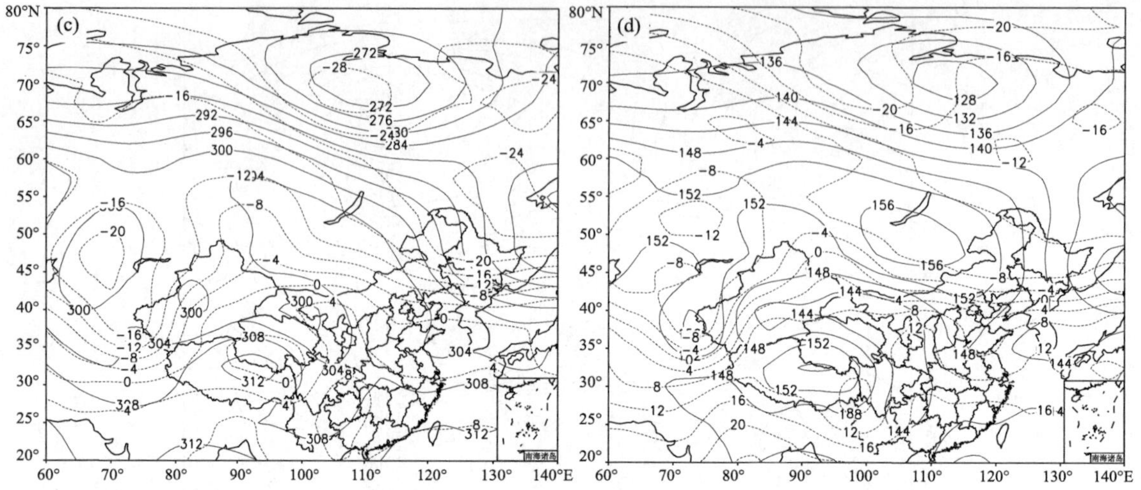

图 8　1994 年 4 月 7 日 08 时高空形势图，实线为位势高度，虚线为温度，其中，a、b、c、d 分布为 250、500、700、850 hPa

5.2　动力、热力结构特征

图 9 为穿过沙尘暴区的经向垂直剖面图，可以看到图中对流层低层形成一明显的垂直次级环流，表现为在 39°N 附近(内蒙古西北沙尘暴发生区)气流辐合上升，到达 700 hPa 后变为水平并折而向北，在 48°N 附近转为下沉，到达 800 hPa 以下又折而向南，形成垂直正环流。其中，辐合上升区刚好位于沙尘暴发生区。另外，在该垂直环流上方，对流层中层 700～250 hPa 之间，南北风在 45°N 附近形成明显的气流交汇，与准静止锋对应。从物理量垂直分布来看，在 39°N 对流层低层辐合区上方，对流层中层准静止锋区南侧为大范围的上升运动区，中心位于 500～600 hPa 之间，而准静止锋北侧则为大范围下沉运动区。从散度垂直分布来看，准静止锋南侧散度分布与上升运动特征较吻合，即低层辐合区位于 39°N 附近 700 hPa 以下，高层辐散区则位于 250 hPa 附近。另外，48°N 附近对流层低层为强辐散区，与垂直次级环流的分布相吻合。从相对湿度来看，在对流层中层，准静止锋区附近为相对湿度大值区，该区域尽管为上升运动区，但并不是最大区，而在 700 hPa 以下对流层低层，在 39°N 附近相对湿度同样较大，但同样不是最大区。

从图 10 穿过沙尘暴区的纬向剖面图可见，在 100°E 与 125°E 间存在一尺度较大的垂直环流，环流中心位于 700 hPa，其下方偏东气流在 100°—115°E 间(发生沙尘暴的区域)向西爬升，在 650 hPa 附近折而向东并继续爬升，上升运动达到 500 hPa 以上，下沉补偿支则出现在115°E 以东地区。可见，与沙尘暴区对应的这两个次级环流表明地面沙尘粒子能够在偏南风、偏东风乃至偏西风的作用下逐步抬升至 500 hPa 以上。另外，从全风速分布可以看到，在沙尘暴区，即 100°—110°E、39°—45°N 范围内，850 hPa 存在一大于 16 m/s的边界层急流区，风速明显强于 500～800 hPa，该急流为沙尘暴的形成提供了有利大风条件。另外，可以看到，在沙尘暴发生区高空急流位于 250 hPa 附近，并未出现急流的下落。

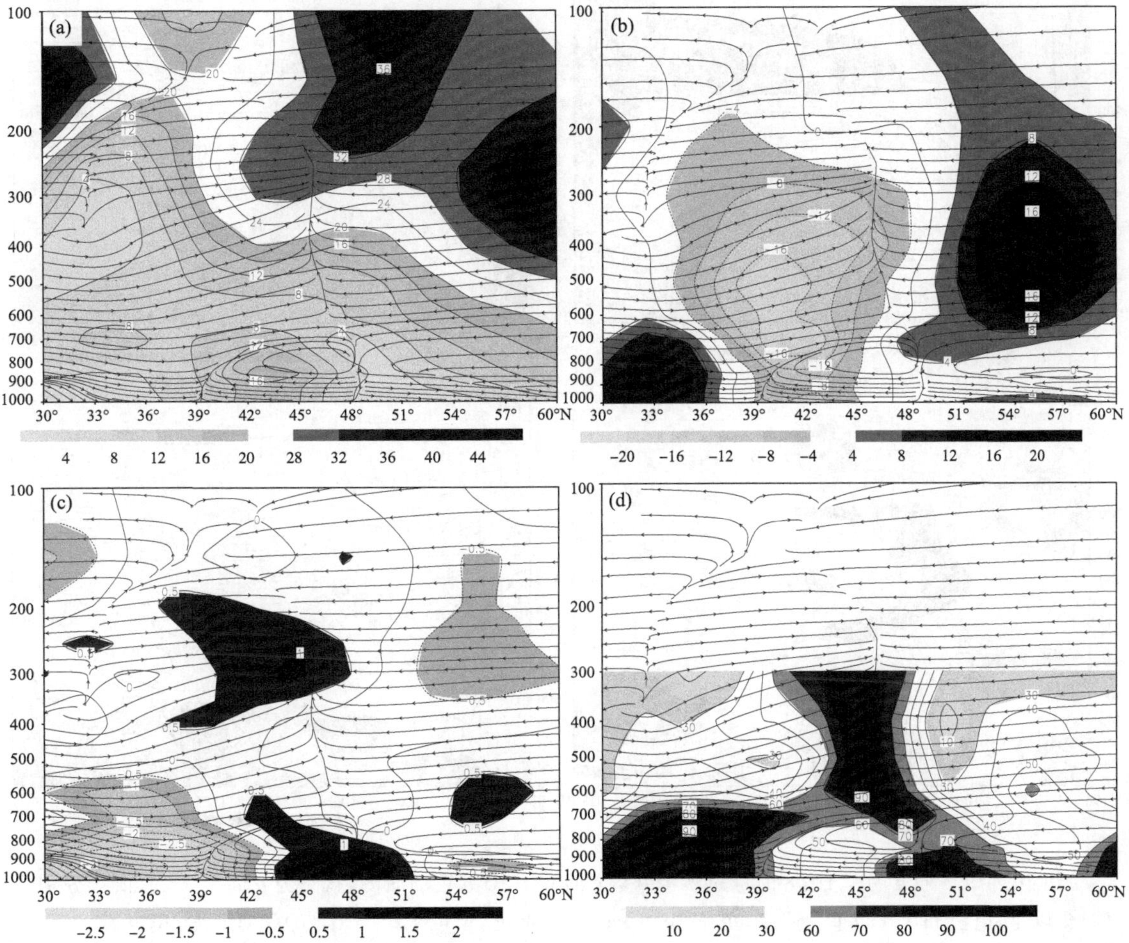

图 9　1994 年 4 月 7 日 08 时穿过沙尘暴区(104°E)经向垂直剖面图,风矢为 v、ω 合成,等值线分别为(a)全风速,单位 m/s,(b)垂直速度,单位 10^{-1} Pa/s,(c)散度,单位 10^{-5} s^{-1},(d)相对湿度,单位%

从物理量垂直分布可以看到,强上升运动区域正好位于发生沙尘暴的区域(100°—105°E)上方,对流层中低层最强,中心位于 650 hPa 左右,强度大于 0.2 Pa/s。从散度分布来看,发生沙尘暴的区域(100°—105°E)出现很强的低层辐合,中心位于 900 hPa 附近,地面倒槽发展是该辐合中心的形成原因,相对而言,这一区域 600 hPa 以上为辐散区,但强度很弱,表明上升运动的触发原因主要是低层辐合导致的。从相对湿度分布来看,发生沙尘暴区域的行星边界层相对湿度在 50%～60%之间,其上方 600～800 hPa 区间,即强上升运动区达到 60%～70%。相对湿度明显高于前述冷锋和蒙古气旋沙尘暴过程,这也是这类过程在地面倒槽偏南侧区域常常出现降水的原因。

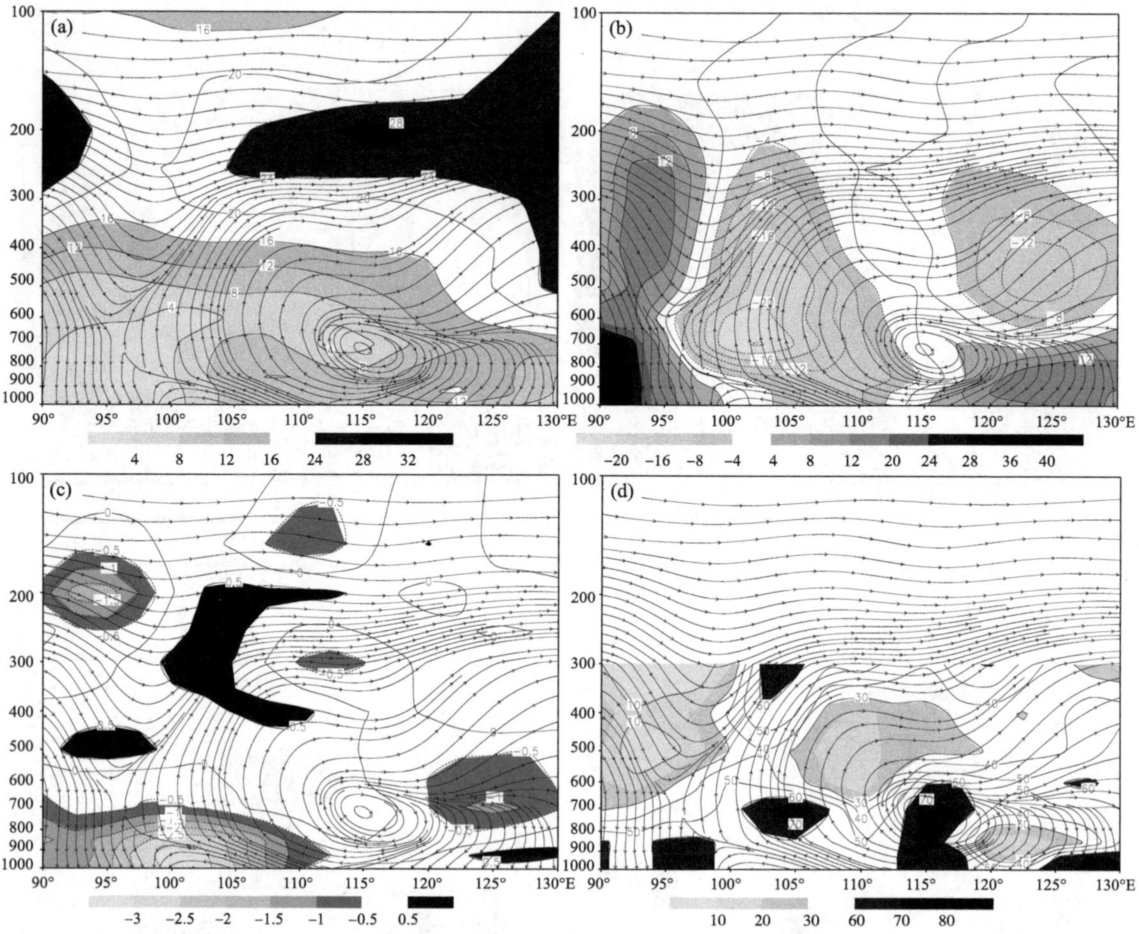

图 10　1994 年 4 月 7 日 08 时穿过沙尘暴区纬向(41°N)垂直剖面图,风矢为 u、ω 合成,等值线分别为(a)全风速,单位 m/s,(b)垂直速度,单位 10^{-1} Pa/s,(c)散度,单位 10^{-5} s^{-1},(d)相对湿度,单位%

6　三类过程的对比

为了更清楚地揭示三类过程环流特征及动力热力结构差异,表 1 给出了三类过程的对比:

表 1　冷锋型、蒙古气旋型、高压底部倒槽型沙尘暴过程对比

类型 特征	冷锋型	蒙古气旋型	高压底部倒槽型
大尺度环流	①中高纬涡(槽) ②中纬度低槽分离 ③强斜压作用	①中高纬槽(涡) ②中纬度低槽分离,形成切断低涡 ③强斜压作用;冷暖锋	①中高纬西北气流 ②中纬度高、低值系统相向发展 ③中纬度准阻塞形势 ④准静止锋
动力热力结构	①系统出现在对流层大部 ②高空急流下落 ③锋区陡立,明显锋面次级环流 ④对流层大部倾斜上升气流 ⑤湿度小	①系统出现在对流层大部 ②明显高空急流下落 ③气旋波 ④垂直上升气柱,贯穿对流层大部 ⑤上湿下干	①对流层中低层系统较强 ②无高空急流下落 ③经向垂直环流 ④沙尘区南侧湿度大

续表

类型 / 特征	冷锋型	蒙古气旋型	高压底部倒槽型
预报着眼点	①天气尺度 ②大范围 ③强沙尘暴 ④持续 1 天 ⑤东南向移动为主强沙尘暴 ⑥地面、对流层低层偏西、西北气流 ⑦无切断低涡	①天气尺度 ②大范围 ③强沙尘暴 ④持续 1—2 天 ⑤东、东南、东北向移动强沙尘暴 ⑥地面、对流层低层偏西、西北、西南气流 ⑦切断低涡	①次天气尺度 ②局地 ③沙尘暴 ④持续 2—3 天 ⑤稳定少动局地沙尘暴 ⑥地面、对流层低层偏东气流 ⑦准静止锋

7 小结

利用天气观测资料对冷锋型、蒙古气旋型、蒙古冷高压底部倒槽型等 3 类天气尺度沙尘暴过程的大尺度环流特征、系统动力热力结构特征进行了对比分析，结构表明：

(1)从大尺度环流特征来看，冷锋型过程在高纬地区对流层大部形成尺度较大的低涡，低涡底部有低槽分离并向南发展，在中纬度 40°N 诱发冷锋生成和发展。蒙古气旋过程中，高纬地区对流层大部为大槽，大槽底部低槽分离并向南发展，在中纬度对流层中低层(500 hPa 以下)形成切断低涡，诱发气旋发展。蒙古冷高压底部倒槽型过程中，中高纬地区受脊前西北气流控制，对流层低层和地面附近蒙古冷高压南压与向北发展的倒槽在中纬度附近遭遇形成准静止锋，并在静止锋南侧诱发沙尘暴。

(2)从动力、热力结构来看，冷锋型过程锋区异常陡立，在 700 hPa 以下近于垂直，锋面抬升呈现梯次分布，上升运动中心分别位于 750 hPa 和 500 hPa，并随高度向冷区倾斜，锋面次级环流明显。蒙古气旋过程中，在气旋区形成达 8～10 个纬距的上升气柱，贯穿整个对流层，并与正涡度及 350 hPa 辐散和 850 hPa 辐合配合。蒙古冷高压底部倒槽型过程中，沿经向在 700 hPa 以下形成垂直正环流，并在沙尘暴区上升，而 700 hPa 以上则形成南北风的明显交汇。沿纬向 300 hPa 以下形成明显的垂直环流，环流中心位于 700 hPa，700 hPa 以下为一致的偏东气流，并在沙尘暴发生区转为上升气流，上升气流扩展至 300 hPa 以下的几乎整个对流层，中心位于 650 hPa。

(3)通过进一步对比三类过程的发生和演变特点，给出了三类沙尘暴过程的预报着眼点。

参考文献

[1] 岳平，牛生杰，张强等. 民勤一次沙尘暴的观测分析[J]. 高原气象，2008，**27**(2)：401-407.

[2] 陈楠，陈晓光，赵光平等. 宁夏不同强度沙尘暴环流差异特征的对比分析[J]. 高原气象，2006，**25**(4)：680-686.

[3] 王可丽，江灏，吴虹等. 2001 年春季中国北方沙尘暴的环流动力结构分析[J]. 高原气象，2002，**21**(3)：303-308.

[4] 郑新江，徐建芬，罗敬宁等. 1998 年 4 月 14—15 日强沙尘暴过程分析[J]. 高原气象，2001，**20**(2)：180-185.

[5] Brazel A J，Nicking W C. The relationship of weather types to dust storm generation in Arizona[J]. *J.*

Cli. ,1986,**6**(3):255-275.

[6] 王式功:1995,我国西北地区沙尘暴时空分布及成因分析,中国科协技术年会论文集,北京:中国科技出版社.

[7] Pauley P M,Baker N L and Barker E H. An observational study of the"interstate 5"dust storm case[J]. *Bull. Amer. Meteor. Soc.* ,1996,**77**,693-719.

[8] 刘景涛,钱正安,姜学恭,等. 中国北方特强沙尘暴的天气系统分型研究[J]. 高原气象,2004,**23**(4):540-547.

[9] 姜学恭,沈建国,刘景涛等. 导致一例强沙尘暴的若干天气因素的观测和模拟研究[J]. 气象学报,2003,**61**(5):606-620.

[10] 姜学恭,沈建国. 内蒙古两类持续型沙尘暴的天气特征[J]. 气候与环境研究,2006,**11**(6):702-712.

A Comparative Analysis on the characteristics of Circulation and Dynamic Structure in Three Types Sand Dust Storm Processes

JIANG Xuegong　YUN Jingbo

Inner Mongolia Autonomous Regional Meteorological Observatory, Hohhot 010051

Abstract The cold front type, Mongolia cyclone type, and inverted trough type, are dominating weather systems inducing sand dust storm processes in East Asia and the North China. For investigating the difference between them to guide the sand dust storm weather forcast, a comparative analysis on the characteristics of circulation and dynamic structure were conducted by using observational data. The result shown: In the cold front type and the Mongolia cyclone type processes, the planetary-scale trough(low) formed at the high-latitude area. The synoptic scale trough(low) separated from it and developed southwards inducing the development of cold front(Mongolia cyclone) at mid-latitude area. The difference between them was a cut low formed at the mid-lower troposphere in the Mongolia cyclone type process. In the inverted trough type process, the northwest wind stream was dominant at the mid-high latitude area. The Mongolia cold high moved southwards and encountered the northwards developed inverted trough and induced a quasi-stationary front formed there. The sand dust storm appeared at south of the quasi-stationary front. In both of the cold front and Mongolia cyclone processes, a typical baroclinic structure formed at the mid-lower troposphere. In the cold front process, the cold front zone was nearly perpendicular to the surface below 700 hPa. A slope ascending air stream appeared and formed a center at the mid-lower troposphere. In the Mongolia cyclone process, an perpendicular ascending air stream column formed at the cyclone area and spread to 8—10 latitudes horizontally and to the whole troposphere vertically with the positive vorticity. In the inverted trough process, along the meridional direction an evident intersection area between the southern and northern wind appeared above 700 hPa. Below it, a positive circulation formed with its ascending branch locating at the dust storm area while its down sinking branch appearing at the cold high area. Based discussed above and a furthermore contradistinction of the development and evolution of the three processes, a forecasts focus was generalized.

Key words: dust storm, weather system, atmospheric circulation, dynamic and thermal structure, comparative analysis

一次蒙古气旋型强沙尘暴过程的数值模拟试验*

姜学恭[1,2]　赵琳娜[2]

1. 中国气象局国家气象中心，北京　100081；2. 内蒙古自治区气象台，呼和浩特　010051

摘　要：本文利用沙尘数值预报模式对一次典型的蒙古气旋沙尘暴过程进行了数值模拟试验，结果表明：模式对沙尘区范围、强度及蒙古气旋不同发展阶段沙尘暴的强度、位置和演变进行了成功的模拟。本次过程沙尘主要源于蒙古国南部及内蒙古中、西部荒漠化地区，沙尘以细尘粒为主（$d \leqslant 22\ \mu m$）。模式模拟的沙尘暴日变化强于实况，其原因是模式没有考虑沙尘气溶胶辐射反馈机制。沙尘气溶胶的辐射强迫将削弱大气层结的日变化，从而有利于沙尘暴强度的维持。

关键词：沙尘暴，数值模拟，层结，沙尘辐射

1　引言

沙尘暴作为一种灾害性天气，其危害程度随着国民经济的发展和人民生活水平的提高而增强。特别是2000年前后，东亚地区沙尘暴的发生出现明显增多的趋势，引起该地区乃至全球的广泛关注。人们对沙尘暴的气候特征及影响因子，沙尘天气的地基、遥感监测技术，沙尘气溶胶物理化学特征及环境、生态效应，沙尘气溶胶辐射强迫和气候效应，气候变化和人类活动对沙尘气溶胶的影响，沙尘起动、输送、沉降等动力学问题，沙尘天气的预测技术等方面进行了广泛的研究工作。[1~6]

提高沙尘暴的预报、预测技术是减轻沙尘暴危害的直接途径，也是开展沙尘暴研究的主要目的。在沙尘暴的预报技术中，沙尘暴的数值预报技术是沙尘暴预报的重要手段。由于沙尘暴是包含大气圈、岩石圈、生物圈等多个圈层及其相互作用的复杂物理过程，下垫面特征对沙尘暴的发生具有重要的影响，同时，风与沙尘粒子的相互作用导致沙尘起动的机理也没有充分的掌握。这使得沙尘暴的数值预报技术远不及降水等其他数值预报技术成熟。在国内，Wang等[7]采用天气模型、摩擦速度和土壤湿度三个控制因子设计了东亚地区起沙机制模型。程丛兰等[8]利用参数化方法及基于MM5开发了沙尘数值预报模式并应用于沙尘天气预报中。孙建华、赵琳娜等[9]基于Shao[10]提出的考虑了微观物理机制的非参数化起沙模型，建立了耦合MM5、起沙模式、沙尘传输模式、GIS（地理信息系统）数据集的沙尘数值预报系统，是更具有理论和实际意义的探索。本文拟利用该模式对2001年4月6—8日强沙尘暴过程进行模拟试验，在检验模式对沙尘暴模拟能力的基础上，揭示一些沙尘暴发生、发展特点。

2　模式简介

沙尘数值预报系统[9]耦合了中尺度天气模式MM5、起沙模式、沙尘传输模式、GIS（地理信息

* 资助课题：国家自然科学基金（40565003）。

系统)数据集(图 1)。图中虚线框内的天气模式、起沙模式和输送模式耦合在一起,同步进行积分。天气模式分别提供摩擦速度、表层土壤湿度给起沙模式,起沙模式计算起沙率及垂直沙通量给输送模式,输送模式利用 MM5 提供的水平风场驱动沙尘,在每个格点计算瞬时沙尘浓度。

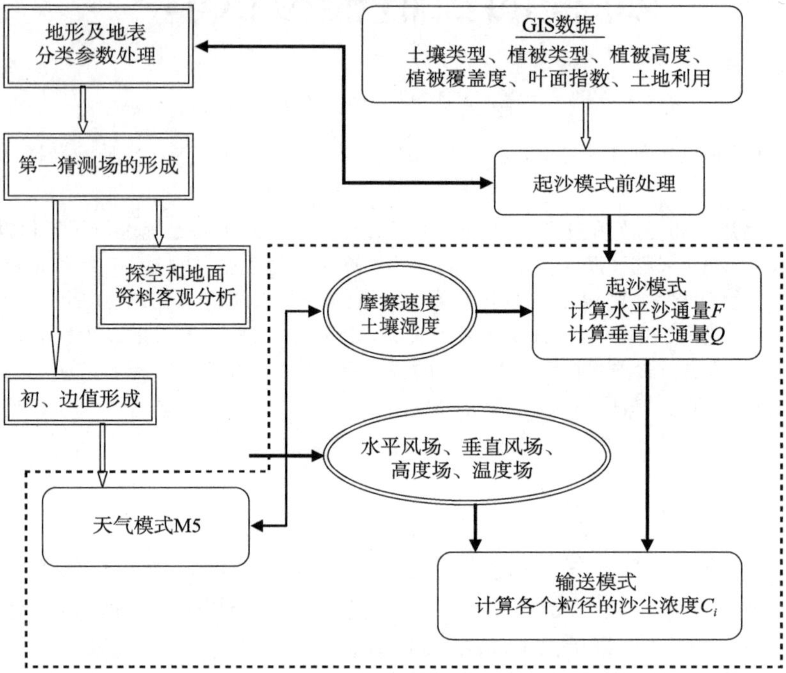

图 1　沙尘天气数值预测系统流程

模式首先计算摩擦速度和临界摩擦速度,当摩擦速度大于临界摩擦速度时,开始起沙。计算起沙量时,将沙尘按粒径分为 6 组($d \leqslant 2$ μm,2 μm$<d \leqslant 11$ μm,11 μm$<d \leqslant 22$ μm,22 μm$<d \leqslant 52$ μm,52 μm$<d \leqslant 90$ μm,90 μm$<d \leqslant 125$ μm),对每一粒径组分别计算水平跃移沙通量和垂直沙通量。

起沙模式需要的陆面参数为土壤类型、土地利用、植被类型、植被高度、叶面面积指数和植被覆盖度。其中土壤类型和植被类型数据国外部分取自美国地质勘察资料(USGS),国内部分取自中国 1∶400 万土壤图数据资料,其他植被资料由遥感数据转化而来。所有数据的分辨率为 $3' \times 3'$。

模式初试场和侧边界采用 NECP 再分析资料。模拟水平分辨率 45 km,模拟区域大致为(25°—60°N,80°—140°E),模拟时段为 2001 年 4 月 6 日 08 时—8 日 08 时(北京时)。

3　天气过程概述

2001 年 4 月 6—8 日沙尘暴过程是一次典型的蒙古气旋型沙尘暴过程,其影响范围包括新疆、甘肃、宁夏、内蒙古、河北、黑龙江、吉林等多个省区,并在内蒙古中部偏北地区引发黑风暴。其造成的灾害,仅在内蒙古锡林郭勒盟北部,导致幼畜死亡 30783 头,走失 3 万头,倒塌和损坏棚圈 4310 间,埋没草场 3000 亩*[11]

* 1 亩=1/15 hm^2。

本次过程沙尘天气的发生与蒙古气旋的发展密切相关。5 日 20 时前，蒙古国有低压产生，该区域个别测站出现了扬沙。6 日 02 时，蒙古气旋形成，气旋冷锋侵入我国新疆，产生小范围扬沙天气。6 日 08 时后，气旋进入快速发展阶段，冷锋东移侵入内蒙古，首先在内蒙古西部引发大范围扬沙天气。14 时，随着气旋的加强，内蒙古西、中部的大部分地区产生沙尘暴和特强沙尘暴，水平能见度迅速下降到 200 m 到 300 m。6 日 20 时—7 日 02 时，沙尘暴东移并在内蒙古中部偏北地区达到最强，水平能见度下降到 200 m 以下（6 日 16 时到 7 日 02 时，内蒙古中部二连浩特、满都拉、朱日和、阿巴嘎、苏尼特右旗等测站能见度降至 0 m）。7 日 08 时后，气旋锢囚并开始减弱，沙尘暴区向东扩展，同时强度减弱。7 日 20 沙尘暴逐渐减弱停止。可见，本次过程强沙尘暴（能见度小于 200 m）主要发生在内蒙古中部地区，呈东西带状。黑风暴主要发生在内蒙古中部偏北地区。

4 结果分析

对于本次过程沙尘暴发生、发展机制，已有一些天气学方面的研究[11]，本文在对沙尘数值预报模式模拟效果进行检验的同时，也希望利用模拟结果对本次过程的特点进行进一步的讨论。

（1）气旋强度和沙尘浓度的模拟

表 1 实况和模拟的气旋中心海平面气压对比（hPa）

时间	6 日 08 时	11 时	14 时	17 时	20 时	23 时	02 时	05 时	7 日 08 时
实况	992	992	987	982	984		977	979	979
模拟	992	989	989	988	987		982	985	985

通过表 1 实况和模拟的气旋中心气压变化可以看出，尽管模拟气旋强度弱于实况，但模式总体上较好地模拟出了蒙古气旋的强度演变，特别是 6 日 17 时之后气旋快速发展过程。另外，模式模拟的气旋中心的位置和移动过程也基本上与实况吻合。

图 2 为模拟的地面沙尘浓度与地面沙尘天气实况的对比，图中显示模式较好地模拟出了本次沙尘天气的出现区域，14 时，模拟沙尘暴区主要在蒙古国南部（阿尔泰—萨彦岭山地南侧和东南侧），国内部分稍弱，位于内蒙古西部荒漠化地区，与实况的沙尘暴天气区分布较为一

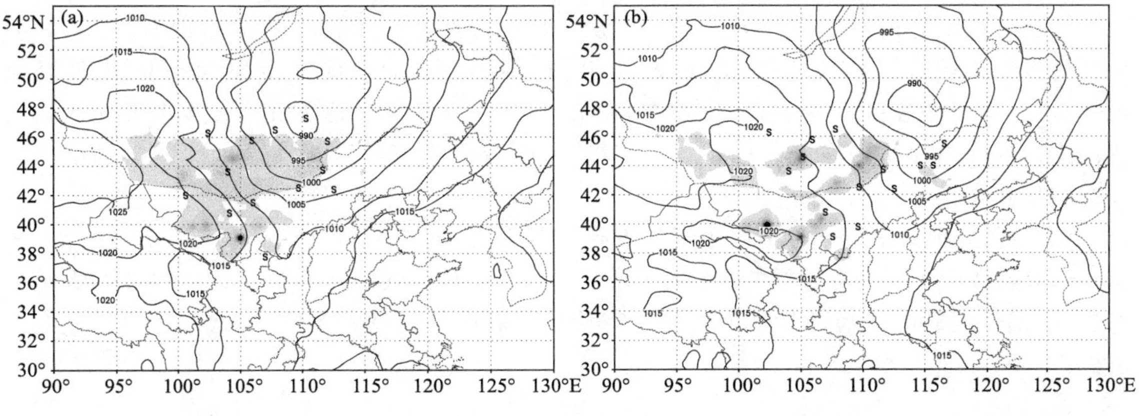

图 2 模拟 2004 年 4 月 6 日(a)14 时和(b)20 时沙尘浓度。阴影区为沙尘浓度，S 为实况沙尘暴出现测站，等值线为海平面气压(hPa)

致。上述地区也是我国北方沙尘暴的两个主要沙源地区。20 时，观测显示蒙古国范围内的沙尘暴减弱，而在我国内蒙古中部偏北地区沙尘暴持续并有所增强。模拟的沙尘高浓度区基本上覆盖了这些地区，但模拟的沙尘浓度弱于实况。总体上，在上面提到的蒙古国南部和我国内蒙古西部荒漠化地区，模拟结果与实况较为吻合，而在内蒙古中部偏北地区，模拟结果要弱于实况。对这一现象下文将进一步讨论。

(2)沙尘区与蒙古气旋的关系

在蒙古气旋旺盛发展阶段，从图 2 模拟沙尘区的分布可以看到，14 时和 20 时沙尘区基本上位于地面气旋冷锋之后。在蒙古气旋发展成熟阶段，从 7 日 02 时沙尘分布(图 3a)可以看到，沙尘区位于冷锋之后且主要集中在气旋副冷锋之后西北气流强盛的区域内。到 7 日 08 时，气旋形成锢囚，沙尘区则位于锢囚锋的冷支之后。可见，在气旋型沙尘暴过程中，沙尘区主要位于气旋冷锋后部，这是由于在气旋冷锋后部对应着强地面风，同时，远距离输送的沙尘也在气旋冷锋后部的下沉运动中形成沉降[11]。

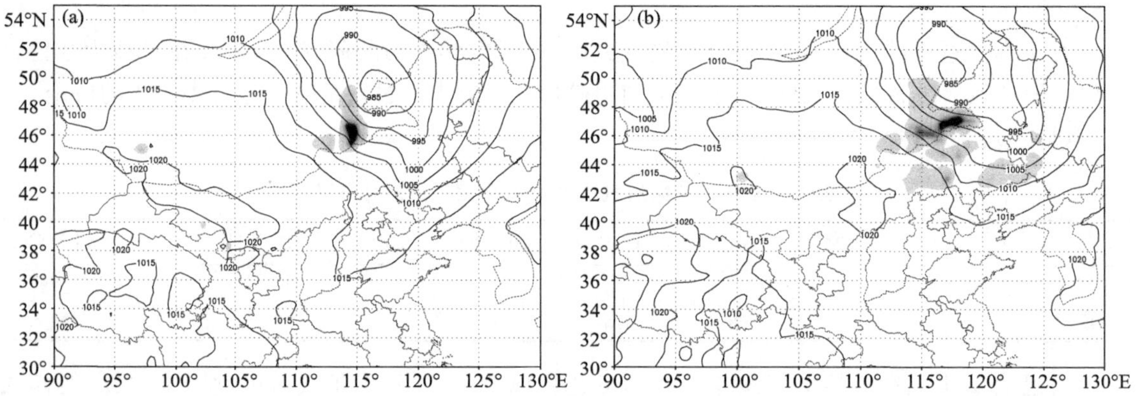

图 3　模拟 2004 年 4 月 7 日(a)02 时和(b)08 时沙尘浓度。阴影区为沙尘浓度，等值线为海平面气压(hPa)

需要注意的是，本次过程 4 月 6 日 14 时前后，在气旋冷锋前部有一小范围的沙尘区出现(图略，可参考文献 11)，诊断分析表明该沙尘区的形成不同于 93 年 5 月 5 日由干飑线引发的沙尘暴，而是由于强烈的地面加热产生了热对流，同时，气旋前部低层西南气流导致局地起沙造成的[11]。图 2a 中在冷锋前同样模拟出了小范围沙尘区。从图 4 穿过冷锋的垂直剖面图更可以清楚的看到这一点。图中，东经 110 度附近为等位温线的东西分界线，表明为气旋冷锋。同时，冷锋后沙尘浓度较大，上扬程度较高，而锋前沙尘浓度较小，上扬高度也较低(该区域海拔高度在 1000 m 左右，即 900 hPa)。这与实况锋面前后沙尘天气的强度和沙尘高度的对比是一致的。可见，模式对于气旋冷锋前部热力扰动产生的沙尘暴也有比较好的反映。

综上可见，模式较好地模拟出了气旋不同发展阶段、气旋不同位置沙尘区的分布特点，表明模式对强沙尘天气具有较好的预报能力。另外，不论对于气旋冷锋后部天气尺度的大范围沙尘区，或是气旋冷锋前部暖区内由热力扰动导致的中尺度的沙尘区，模式都能给出较好的模拟。

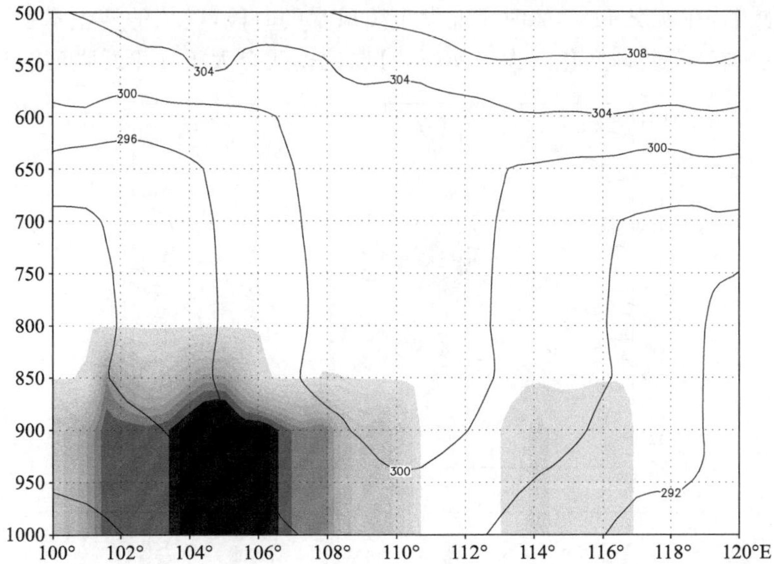

图 4 4 月 6 日 14 时穿过气旋冷锋东西向垂直剖面图。阴影区为沙尘浓度，等值线为位温(K)

(3)沙尘源地和起沙

在模拟结果与实况吻合较好的前提下，模拟结果可以揭示一些沙尘粒子的粒径分布特征。图 5 为 4 月 6 日 14 时起沙最为强盛时段模拟的不同粒径沙尘的起沙量。其中，a 为黏土($d\leqslant 2\ \mu$m)，b 为细粉土($2<d\leqslant 11\ \mu$m)，c 为中粉土($11<d\leqslant 22\ \mu$m)。可以看到导致本次沙尘暴过程的沙尘粒子主要为尘粒，大于 22 μm 的沙尘粒子很少(图略)。其源地主要位于蒙古国(西)南部戈壁、我国内蒙古西部巴丹吉林沙漠、腾格里沙漠等以及位于锡林郭勒草原中部的浑善达克沙地。另外一个显著特点是在蒙古国南部，三种粒径的粒子分布比较均匀，而在我国内蒙古中西部沙源地，粒径主要为细粉土($2\ \mu\text{m}<d\leqslant 11\ \mu$m)。而这些地区都是荒漠化程度较高的沙漠地区。因此，就本次过程而言，荒漠化地区是沙尘暴的主要供尘区。另外，比较对沙尘暴的总体贡献，蒙古国南部的戈壁沙漠显然较国内内蒙古中西部荒漠化地区的贡献更为显著。

(4)沙尘天气的日变化

对比图 2 和图 3 可以看出，随着时间推移，从白天到傍晚乃至深夜，模式模拟的起沙量迅速减小，即 14 时起沙量最大，之后逐渐减小，到 02 时达到最小，到次日 08 时起沙量又明显增强。比较沙尘浓度，也出现同样的情况。可见，模拟的沙尘强度出现很强的日变化。

沙尘暴普遍具有明显的日变化，这是由于地面加热的日变化所致：白天地面加热使大气层结不稳定性增强，而夜间则因为辐射冷却形成逆温稳定层。层结越不稳定，则越有利于沙尘的扬升和维持，沙尘暴也就越强。但同时，决定沙尘暴强度的直接因素是天气系统的强度，即本次过程中的蒙古气旋。天气系统的强度演变取决于环流系统的演变，并不严格遵循日变化(尽管也受到日变化影响)。就本次过程而言，蒙古气旋在白天开始增强，傍晚至午夜是气旋爆发性发展的阶段[11]。实况观测到的沙尘暴强度变化与气旋强度的这一变化是一致的。图 6 为本次过程沙尘暴最强区域—内蒙古中部偏北地区二连浩特站(发生黑风暴)6 日 08 时—7 日 08 时地面气象要素演变，可以看出，6 日 14 时开始出现沙尘暴(能见度 800 m)，之后，沙尘暴继续增强，能见度持续减小，到 7 日 02 时能见度降为 0 m，出现黑风暴。之后到 7 日 08 时，沙尘暴强度才开始缓慢减弱。而图 7 模拟的二连浩特沙尘暴强度演变是：最强沙尘暴出现在 6 日 17 时左右，之后即迅速减弱，在实况 11—02 时实况发生黑风暴的时段，模拟的沙尘暴强度已经很弱。可见，模式模拟的沙尘暴强度日变化远大于实况，可以说模式夸大了沙尘暴强度的日变化。

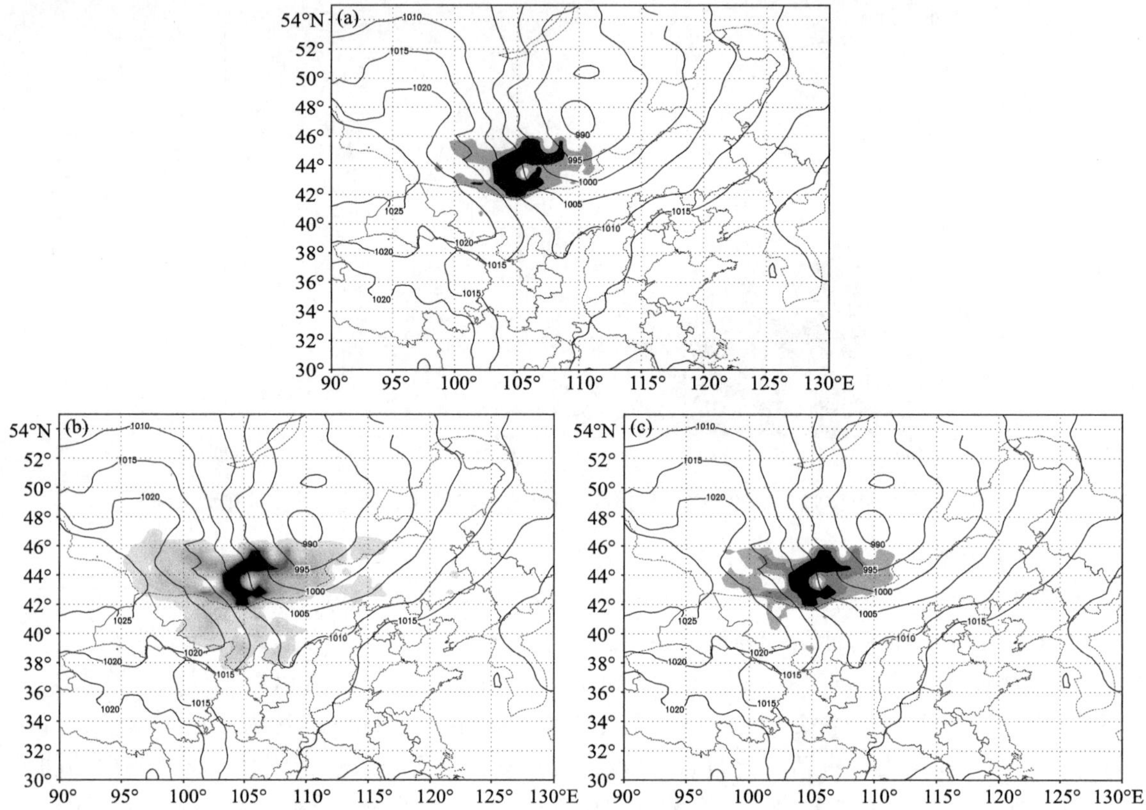

图 5　模拟 4 月 6 日 14 时垂直尘通量($d \leqslant 2$ μm)和海平面气压(等值线)

(a. $d \leqslant 2$ μm；b. 2 μm$< d \leqslant 11$ μm；c. 11 μm$< d \leqslant 22$ μm)

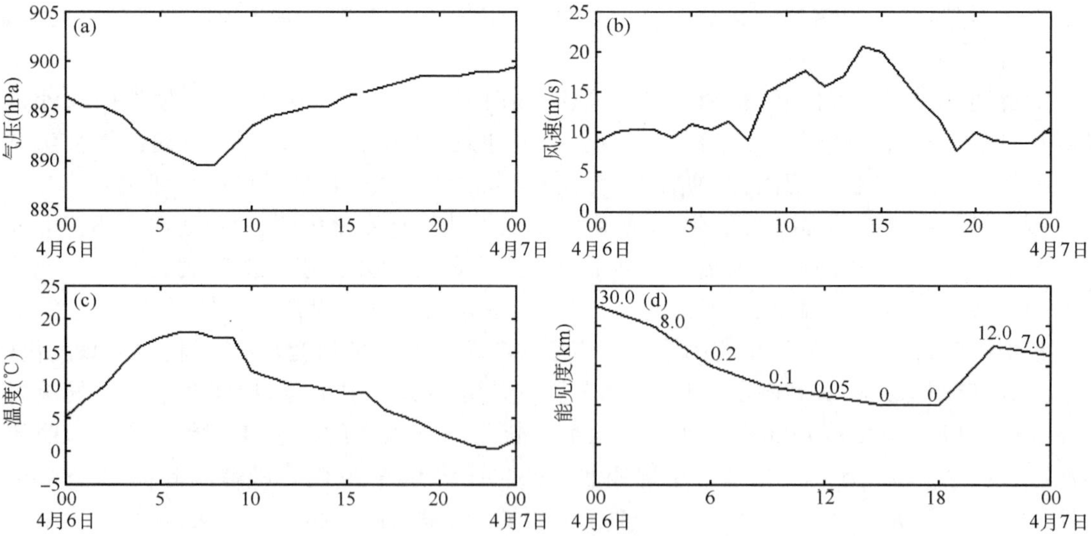

图 6　二连浩特 4 月 6 日 08 时至 7 日 08 时(图中为世界时)地面气象要素变化

(a)气压，单位：hPa；(b)风速，单位：m/s；(c)温度，单位：℃(d)能见度，单位：100 m

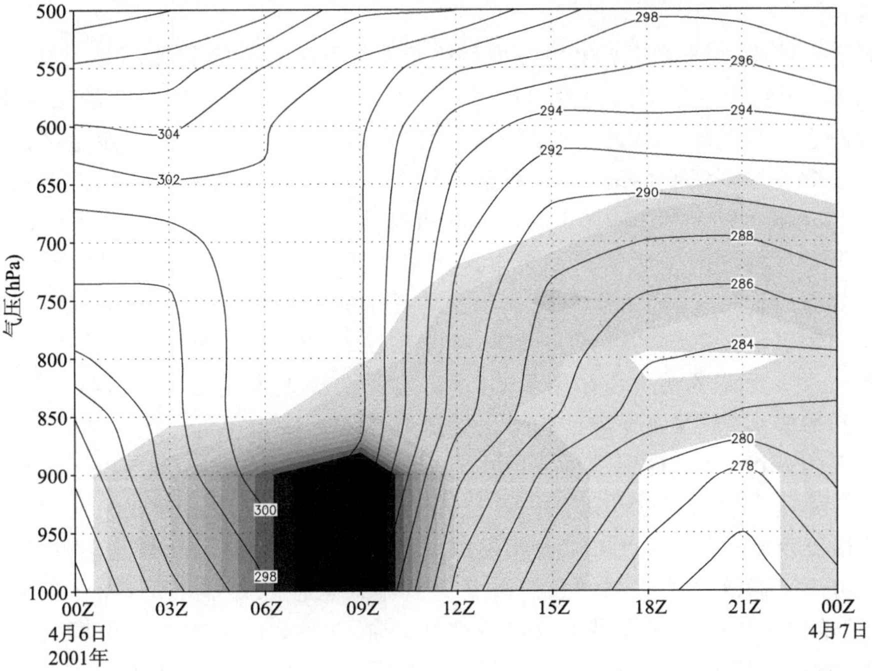

图 7 模拟 4 月 6 日 08 时—7 日 08 时二连浩特附近沙尘浓度和位温演变

导致沙尘暴日变化的直接因素是地面加热的日变化,因此,削弱沙尘暴日变化的因素可能也是通过影响地面加热实现的。对于沙尘暴过程而言,这一因素是存在的,那就是沙尘的辐射强迫。陈受钧[12]等研究表明,沙尘气溶胶的辐射强迫能够使对流层中的沙尘层形成等熵层结(混合层)。Carson 等[13-14]研究表明,在撒哈拉沙尘远距离传输过程中,沙尘层表现为等熵混合层。就本次过程沙尘层位于边界层中的情况,在白天,沙尘层的遮盖将削弱地面接收的太阳辐射(长波辐射),同时沙尘粒子将吸收太阳短波辐射并加热大气,结果是削弱大气低层的热力不稳定而使层结接近中性,即混合层;夜间则相反,沙尘层的遮盖削弱地面的辐射冷却,同时通过向外的长波辐射冷却低层的沙尘层大气。产生的结果也是削弱逆温层结而出现中性混合层。即沙尘层的辐射反馈机制将削弱大气层结的日变化,使大气层结更接近中性。一些研究也证明了这一现象的存在[15]。而层结日变化的削弱无疑将导致沙尘暴强度日变化的减小。

为了了解实际大气层结的日变化情况,下面仅利用地面位温进行一下简单的分析。6 日 20 时,根据观测资料计算的二连浩特 850 hPa 位温为 293.5 K,地面(大约 895 hPa)位温为 293.4 K,表明该层内为等熵混合层。而从图 7 可以看到模拟的 850 hPa 和地面(900 hPa)位温分别为 291 K 和 289 K,为弱稳定层结,可以看出在傍晚前后模拟结果部分夸大了大气的层结稳定度。假设这种夸大是由沙尘辐射导致的,那么考虑沙尘辐射时的理想模拟结果是 850 hPa 和地面的位温均为 290 K 左右,与 293.5 K 的差别 3.5 K 为模式误差。7 日 02 时(实况黑风暴在维持),模拟的 850 hPa 和地面位温分别为 283 K 和 279.5 K,变为明显的稳定层结。实况 850 hPa 位温无法得到,就以模拟 850 hPa 位温加上模式误差 3.5 K 来保守估计,则约为 286.5 K,实况地面(899 hPa)位温为 285 K,因此,依然可以近视看作混合层(<3.5 K/1 km)。可见,由于没有考虑沙尘的辐射反馈,模式明显夸大了夜间大气的层结稳定度。

(5)小结

模式除了对沙尘暴强度日变化的模拟偏强之外,还存在一些其他不足。从图 2,3 模拟 6 日 14 时—7 日 08 时沙尘浓度的演变可以发现,模拟沙尘区似乎多集中在沙源地附近,沙尘的远距离传输似乎不很明显。从遥感监测图像(略)可以地看到 7 日 08 时沙尘区沿气旋前部气旋性弯曲一直被输送至气旋西北象限,而模拟沙尘区仅仅出现在冷锋之后。但从图 7 可以清楚地看出在 800 hPa 以下和 800～650 hPa 间有两个沙尘传输层存在。表明模式沙尘传输过程存在一些不足。其原因可能是多方面的。本文在此仅讨论一种因素的影响,那就是层结因素。正如上节所讨论的,真实大气情况下沙尘的辐射强迫能够使沙尘层形成等熵层结。而在没有考虑沙尘辐射反馈影响的模式大气中,沙尘层变为非常稳定的层结。稳定层结无疑较等熵层结更有利于沙尘粒子在重力作用下的沉降。因此,模式大气中,沙尘粒子上扬后可能较真实大气更快的形成沉降,因而导致沙尘输送的距离小于实际情况。

综合上面的分析,可以看到总体上模式模拟的沙尘区范围、强度与观测记录吻合,也模拟出了蒙古气旋不同发展阶段沙尘暴的强度、位置及演变。沙源地附近观测到的沙尘暴在模式中一定能够得到反映,从这一点可以看出模式的起沙过程的处理是比较理想的,特别是采用了非参数化的考虑了微观物理机制的起沙模型。但模式也存在一些不足,如沙尘传输距离小于实况,为沙尘传输模块的处理提供了改进的空间。需要着重指出的是由于模式没有考虑沙尘的辐射反馈,使得模拟的沙尘暴日变化远强于实际情况。长久以来,人们一直关注沙尘气溶胶的辐射强迫,但多从气候角度研究其对大气的影响。在天气尺度,目前国内的沙尘数值预报模式都没有将沙尘的辐射反馈加以考虑。因此,有必要全面评估沙尘辐射强迫对天气尺度系统影响,在此基础上将沙尘辐射反馈机制引入沙尘数值预报模式中,从而进一步提高沙尘暴的预报能力。

5 结论

本文利用沙尘数值预报模式对 2001 年 4 月 6—8 日蒙古气旋型沙尘暴过程进行了模拟试验,结果表明:

(1)沙尘数值预报模式对这一过程进行了较成功的模拟,模拟的沙尘区范围、强度及蒙古气旋不同发展阶段沙尘暴的强度、位置和演变基本与实况吻合。

(2)不论对于气旋冷锋后部天气尺度的大范围沙尘区,还是冷锋前部暖区内的中尺度热力扰动导致的沙尘暴,模式都能进行较好的模拟。

(3)导致本次沙尘暴过程的沙尘以尘粒为主($\leqslant 22\ \mu m$),主要源于蒙古国南部和内蒙古西、中部荒漠化地区。蒙古国南部沙源的贡献更为明显,沙尘为黏土($d \leqslant 2\ \mu m$)、细粉土($2 < d \leqslant 11\ \mu m$)、中粉土($11 < d \leqslant 22\ \mu m$)。我国内蒙古中西部沙源的贡献相对较小,沙尘以细粉土为主。

(4)模式的不足表现为模拟的沙尘暴日变化强于实况。同时,沙尘传输距离小于实况。前者的主要原因是模式未考虑沙尘气溶胶的辐射反馈机制。沙尘气溶胶的辐射强迫将削弱大气层结的日变化,从而有利于沙尘暴强度的维持。

参考文献

[1] 叶笃正,丑纪范,等.关于我国华北沙尘天气的成因与治理对策.地理学报,2000,**55**:513-521.

[2] 刘景涛,郑明倩. 内蒙古中西部强和特强沙尘暴的气候学特征. 高原气象,2003,**2**(1):51-64.

[3] 郑新江,徐建芬,罗敬宁,等. 1998 年 4 月 14—15 日强沙尘暴过程分析. 高原气象,2001,**20**(2):180-185.

[4] 项续康,江吉喜. 西北地区强沙尘暴成因的中尺度分析. 高原气象,1996,**15**(4):448-455.

[5] 刘景涛,钱正安,姜学恭,等. 中国北方特强沙尘暴的天气系统分型研究. 高原气象,2004,**23**(4):540-547.

[6] 王可丽,江灏,吴虹. 2001 年春季中国北方沙尘暴的环流动力结构分析. 高原气象,2002,**21**(3):303-308.

[7] Wang Z,Ueda H,and Huang M. A deflation model for use in modeling long-range transport of yellow sand overm East Asia. *J. Geophy. Res.* ,2000,**105**(D22):26947-26960.

[8] 程丛兰,王迎春,刘伟东,等. 与动力场相耦合的区域沙尘分档模式及个例模拟. 科学通报,2004,**46**(19):2007-2013.

[9] 孙建华,赵琳娜,赵思雄. 一个适用于我国北方的沙尘天气数值预测系统及其应用试验. 气候与环境研究,2003,**8**(2):125-142.

[10] Shao Y P. A model for mineral dust emission. *J. Geop. Res.* ,2001,**106**(D17):20239-20254.

[11] 姜学恭,沈建国,刘景涛,陈受钧. 导致一例强沙尘暴的若干天气因素的观测和模拟研究. 气象学报,2003,**61**(5):606-620.

[12] Chen S J,Kuo Y H et al. The effect of dust radiative heating on low-level frontogenesis. 1995,*J. Atmos. Sci.* **52**:1414-1420.

[13] Carlson T N,Prospero J M. The large-scale movement of Sahalan air outbreak over the northern equatorial Atlantic. 1972,*J. Appl. Meteo.* **11**:283-297.

[14] Carlson T N,Benjamin S G. Radiative heating rates for Saharan dust. 1980,*J. Atmos. Sci.* **37**:193-213.

[15] 姜学恭,等. 内蒙古两类持续沙尘暴的天气特征. 气候与环境研究,待发表.

A Numerical Simulation on a Severe Sand-Dust Storm Process Caused by the Mongolia Cyclogonesis

JIANG Xuegong[1,2] ZHAO Linna[2]

1. Inner Mongolia Meteorological Observatory,Huhhot,China 010051;

2. National Meteorological Center,China meteorological administration Beijing,China 100081

Abstract By using a dust storm numerical predict model, which coupling MM5, dust raising and transporting model and GIS dataset, a success simulation was conducted on a severe dust storm process occurred at northern China from April 6 to April 8, 2001. The result shown that the strength and evolution of Mongolia cyclone, the intensity, positions and evolution of dust storm have been successfully simulated. The dust source regions in this process were the deserts in Mongolia and western Inner Mongolia. The small diameter dust($d \leqslant 22\ \mu m$)was in a dominant position among the dusts particles contributing to the formation of dust storm. The modeling diurnal change of dust storm strength, which acted as intensive in day time while as weak at night, was more great than the observational one. The source of it is the dust radiation heating, which was not considered in the coupling model, conduced an import feedback forcing on the weather system. It will minimized the diurnal change of atmospheric stratification and so make for the sustain of dust storm.

Key words: Dust storm, Numerical simulation, Stratification, Dust radiation

下编　沙尘远距离传输及其强迫要素的模拟分析

An Observational and Numerical Study on the Topographic Influence on Dust Transportations in East Asia

*

JIANG Xuegong[1]

GAO Tao[2]　CHEN Shoujun[3]

1. Inner Mongolia Autonomous Regional Meteorological Observatory, Hohhot 010051;

2. Inner Mongolia Meteorological Institute, Hohhot 010051;

3. Department of Atmospheric Sciences, the School of Physics, Beijing University, Beijing 100871

Abstract Based on observations and numerical simulation techniques, the topographical impacts on dust transportations in East Asia were studied. According to the results of investigations in dust weathers in East Asia, two areas frequently attacked by dust storms have been confirmed, one is the region includes the western part of Inner Mongolia and the southern region of Mongolia (the region named the Mongolia Plateau), and the other is the Tarim basin. The most frequently dust storm occurrence area of the first one appears in its hinterland while for the second one, the south margin of it is the most frequently happening zone. In addition, there is a sub-frequently dust storm attacked region in the area from the northeastern edge of the Tibetan Plateau to the Loess Plateau. The dust storms frequently occurred in the Mongolia Plateau is not only due to the abundant sand and dust sources, but also based on the special topographic conditions of East Asia as well. Three influence factors have been found to be helpful for dust storm forming in the hinterland of the Mongolia Plateau. The most significant one is the "canyon low level jet" (CLLJ), which is formed around the south areas of Altay-Sayan Mountains with an east-west direction at the beginning, then, it enhances the surface wind speed significantly. Due to the obstruct effects of the CLLJ, a lot of dust particles carried by the southward down slop cold air mass would pile up over the south slop of the Sayan Mountains. Moreover, the uneven surface conditions are very favorable for raising the dust particles into the upper air. With the dust particles pile up continuously in the atmosphere, the dust layer in the troposphere can be recognized as 'dust accumulating container', which provides abundant dust particles for the later transportations to the downstream areas. In spite of this, the topographical features of East Asia also conduct great influence on dust transportations. Generally, the easterly CLLJ enhances the easterly dust transportation. The down slope air current at southern Sayan Mountains and the rounding effect of current at the northeastern edge of the Tibetan Plateau enhance the southerly dust transportation. Considering of weather system influences, the weathers caused by cold

* Sponsored by the NSFC (the National Natural Science Foundation of China) Project under Grant No. 41365003.

fronts frequently appear over the area including Mongolia Republic and North China in springtime. The cold front system, in general, carries the sand and dust to southward. Compared to all topographic influence elements, the rounding effect of current is the strongest one and head to southward. Under the combined influence of the cold front and the rounding effect of current, most part of the transported sand and dust depose in the region of the Loess Plateau.

Key word: Dust transportation, Topographic influence, Observations and numerical simulations

1 Introduction

The frequently occurrences of dust storms in East Asia has been drawing great attention of the people in the world. According to a long term dust storm observation, the dust storm phenomenon appears not only around the dust source regions, but also spreads to the remote downwind regions as well. As a kind of disaster weather, the dust storm seriously affects the local industry, agriculture, traffic, people's health and living environment. According to Shi's primary estimate[1], annually, there is about 1.0—3.0 Gt dust aerosol emit from the earth surface into the atmosphere and about 120 Mt(diameter< 41 μm[2])of the dust particles from Asia. In the troposphere over the zone of the middle latitudes throughout eastern Asia to western North America, the dust aerosols mainly source from Asia which fact is confirmed by means of observation and numerical simulation analyses on spring dust emissions in the studies of Zhang et al.[2,3] and Zhao et al.[4]. In addition, it has been also indicated that about 70% of the Asia dust come from the deserts in Mongolia and Northwest China[2]. 51% of these dust particles deposits to the source regions[4], the left are uplifted into the free atmosphere (3~10 km) by cyclone systems and then transported to the downwind regions along the belt region around 40°N, among which 21% deposit to Asia inland, 9% deposit to the seacoast of Pacific, 16% deposit to the Pacific Ocean and 3% deposit to North America through the trans-Pacific transportations[4]. Moreover, the process of dust trans-Pacific transportation occurred in western Mongolia on May, 1998 has been analyzed in Hacker's study[5].

A dust storm, at the beginning, only influences a relative small area (the source region), but with the dust emission and long distance transportation, the area of its impacted downstream regions may enlarge dramatically. Meanwhile, for the population density, productivity and economic level of the downwind regions are always higher than that of dust source regions, the dust impacts on the inhabitants, climate, environment and ecosystem may increase by a hundred times following the dust transportation. Equally important, the measurement of decreasing the desertification should be seriously considered not only focus on the dust storm itself but also needed to know the rule of dust transportations as well.

The crucial element of dust transportations is an appropriate atmospheric circulation system, which determines the directions, distances and influence areas of the dust transportations[6]. In a deeper view, the more directly controlling power of the transportations is the dominant air current in the dust layer in the free atmosphere, especially over a plain area. For instance, the dusts from the Sahara Desert, in general, are transported by the easterlies above the tropical zone to the Atlantic Ocean. Otherwise, it is different when a dust storm occurred in the areas with a higher altitude and uneven surface, for example, East Asia, in where the dust transportation style is mainly determined by the characteristics of local topography. Since there are many higher mountains in the sand and dust source regions with an altitude over 1000 m on average, the dust particles are always uplifted into the middle-upper layer of the troposphere[7]. Furthermore, following the downwind direction of East Asian dust aerosols transportation path, the altitude decreases gradually from the northwestern region to southeast areas. Under this kind of surface conditions, the East Asian dust aerosols can be easily moved by the northwest current to the large areas in its downwind direction. Sometimes the dust can even be transported to the coast line of the Pacific Ocean or to North America. Few literatures talking about the impacts of the topography on dust storms in East Asia can be found in issued scientific journals. Therefore, some referential analysis results have been presented in this article, including statistically analyzing the historical observational data, particular dust event analysis compared to the outputs of numerical simulations to deduce the potential influence of the topography on dust particle transportations in East Asia.

2 The topography characteristics of eastern Asia and its potential influence on dust storm

The spatial distribution of annual dusty days (including the weather phenomena floating dust, sand blowing and dust storm) of eastern Asia is shown in Fig. 1. The data set used to draw the figure is obtained from China Meteorological Administration (CMA). We give a day with a dust weather score if there is a dust phenomenon recorded within four daily regular observations (00:00, 06:00, 12:00 and 18:00 UTC). The time scale of the data set is from January 1, 1980 to December 31, 1999. There are two dust weather frequent occurrence regions can be found in the figure, one is the area contains southern Mongolia and the western part of Inner Mongolia with 20—50 dusty days on average per year. The other one is located in the Tarim basin of southern part of Xinjiang with 30—100 days on average per year. The two regions are commonly recognized by scientists as the two major sand and dust source regions in East Asia[8].

The topographical features of East Asia can be viewed in Fig. 2. By comparing it to Fig. 1, the difference between dust storm influences of the two frequent regions can be found. Although the dust weather frequency of the second region is higher than that of the first region, the impacts of it are not as serious as that of the first region. Due to the special basin features of the second region, there are just few of dust storms can move out of the basin. Therefore, its impact scope is relatively small. On the other hand, depending on the plateau characteristics of the first region (the Mongolia Plateau) and helped by favorable weather conditions, most of the dust storms can usually make their way to reach the downstream regions, even approach to the eastern shore of the Pacific. Compared to the topographical conditions of both source regions, we can found some similar and dissimilar features. The similar condition is that both of the two source regions are located in relative lower areas comparing to their surrounding terrains. The first one covers large territory between the Sayan Mountains and the Qilian Mountains, and the second one lies in the area between the Tianshan Mountains and the Tibetan Plateau. The dissimilar feature of the two regions is that the second one is surrounded by the Tianshan Mountains in the north, the Pamir Plateau in the west and the Altyn Tagh Mountains in the south. The Hexi Corridor is only the out gate for dust exportation. Due to the special topographic conditions, the uplifted dust was trapped by the surrounding mountains and can scarcely be moved out from the basin. Consequently, tracking the path of the Hexi Carridor, there is a small part of the dust can make severe influence onto its downstream regions. Contrarily, the situation of the first source is different. The Sayan Mountains lies in its North West area and directing from northwestwards to southeastwards while the Qilian Mountains Locates in its southwest region heading from northwest direction to southeast. The particular surface shape with those two mountains makes a bell-mouthed terrain around the first source region. Actually, the wide mouth of the bell-shaped topography opens to its downwind regions (Fig. 2). As a result, the dust particles lifted by upwind and produced by cold air attacks over the Mongolia Plateau are very easy to be transported outside of the source region, heading to the southeastwards. In spite of this, the down slope terrain condition of the large zone which is from the first source area to the downwind region is helpful for dust particle diffusions.

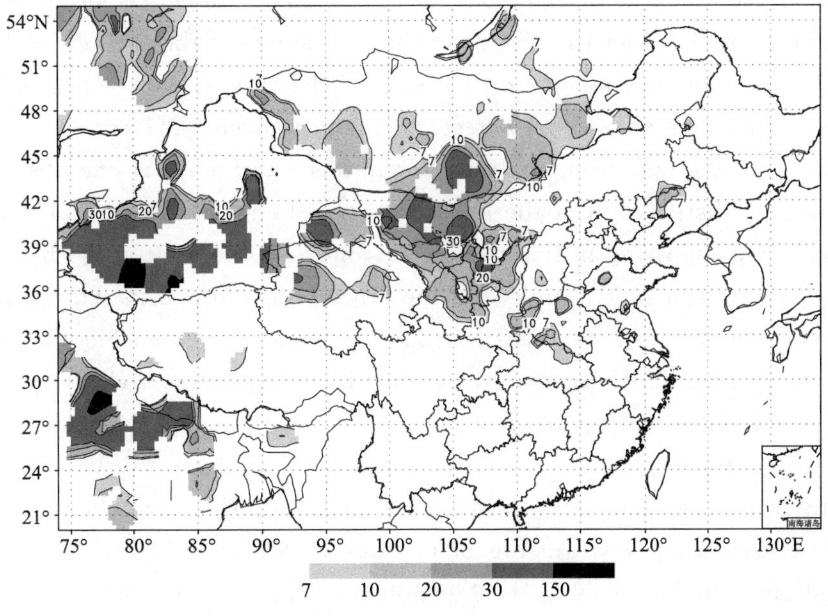

Fig. 1 Spatial distributions of annual mean dust days in East Asia.

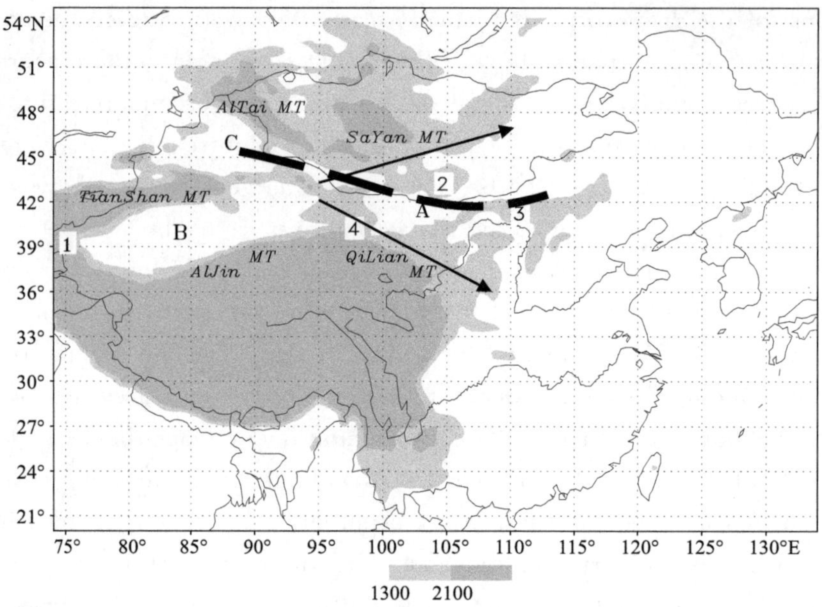

Fig. 2 Topographic conditions of East Asia. (1: the Pamir Plateau; 2: the Gobi Altay Mountains, 3: the Yinshan Mountains; 4: the entrance of Hexi Corridor; A: the desert area from western Mongolia to Inner Mongolia, China; B: the Tarim basin; C: the Junggar basin. Here A and B also indicate the dusty weather frequently happening areas. The arrows act as a bell-mouthed terrain. The thick dashed line represents the canyon).

3 Description of the dust storm case

The bell-shaped topographic condition of the Mongolia Plateau should be a favorable factor for dust spreading. In order to investigate the mechanism of the spreading, the methods of synoptic study and numerical simulation are employed to analyze the topographic effects on dust transportations. Therefore, the severe dust storm event occurred in the days from Mar. 18 to Mar. 22, 2002 has been analyzed, for it was a tremendous one observed in recent years. The intensity distribution of the dust storm can be seen in Fig. 3. In a deep view, we can find two centers appearing in Inner Mongolia, one is in the west area where is dominated mainly by the deserts and sandy lands, and the other can be found in the region which contains the middle area and the southeastern part of Inner Mongolia. The influenced area of those two centers covers more than 90 percent of the dust weather frequently attacked areas in China. Meanwhile, the dusty weather also appeared in Mongolia, the western shore of Pacific such as Korea, Japan et al.

The mentioned serious dust case was mainly caused by the Mongolian cyclone. The previous weather situation is shown in Fig. 3. The dusty weather area can be seen as a volute shape that indicates the influence of the Mongolian cyclone. At 00:00 UTC, Mar. 18, 2002, a trough appeared at middle layer of the troposphere (500 hPa) above the central Asia and deepened gradually along with an eastward movement. When it reached to eastern Xinjiang (95° E) at

00:00 UTC, Mar. 19, 2002, a cyclone formed at the lee side of Sayan Mountains. Meanwhile, abundant sand and dust were blown up into the high sky by the cyclone over western region of Mongolia. At 12:00 UTC, Mar. 19, 2002, associated with deepening trough and the cyclonic cold front of the strengthening Mongolian cyclone on the surface, the dust storm started to invade Inner Mongolia, China. The cyclone became stronger on the next day and the dust storm led by the cyclone swept most part of North China. On Mar. 21, 2002, following the front movement, the dust storm expanded to Jinlin and Liaoning Province. Although the dust storm had disappeared gradually on the local spot with the weakness of the cyclone during the later days, the suspended sand and dust still kept in the high level of the atmosphere and spread continuously to the Western Pacific ocean.

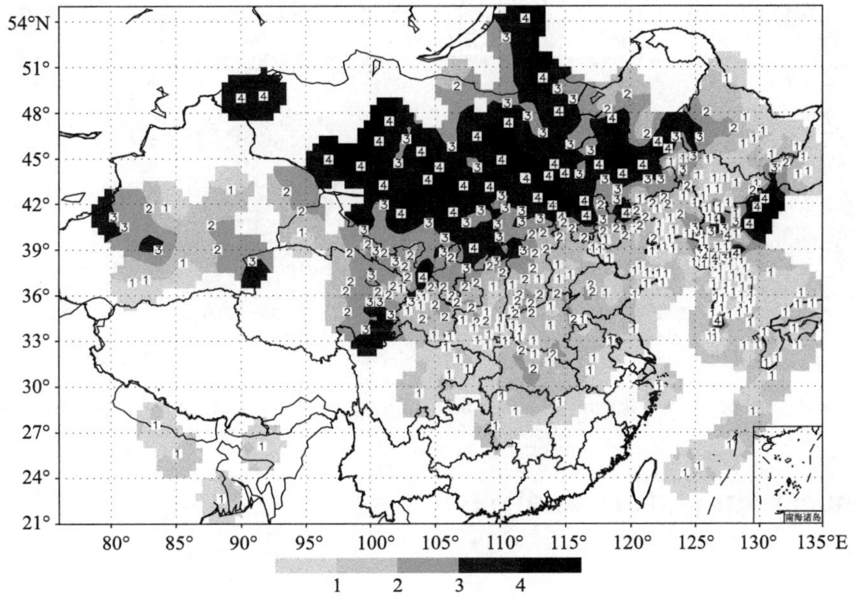

Fig. 3 The distribution of dusty weather in East Asia from Mar. 18 to Mar. 22, 2002. (1, 2, 3, 4 and different darkness denote floating dust, blowing dust, dust storm and severe dust storm, respectively)

4 The model and model set

The numerical model developed in Beijing Institute of Urban Meteorology, CMA was used in this study. It is created by coupling MM5 (the Fifth-Generation NCAR/Penn State Mesoscale Model) and a set of mass conservation equations for the dust particles. In order to distinguish the dust particles precisely, the diameter of the particles are sorted into sixteen degrees. The dust lifting procedure used in the model is the method of parameterization, which is related to surface sorts, canopy and soil moistures. The simulated dust deposition scheme includes dry and wet processes. The concerned crucial dust blowing element is the fricative velocity on the surface[9].

The severe dust event of May 19-22, 2002 is simulated. The grid resolution is 30 km and the grid number is 131×121. The T213 analysis and forecast data obtained from the National Meteorological Center of China, were used as the initial and boundary conditions. The modeling started at 00:00 UTC, Mar. 19, 2002 and lasted for 48 h. The information of real topography was utilized in the control run (Fig. 4a) while in the sensitive test, we adjusted the altitude around the Altay-Sayan Mountains and reduce the highest altitude from about 2500 m to 1500 m. The adjusted topography in sensitive test is presented in Fig. 4b. The following study will focus on the strong impact period (from 00:00 UTC, Mar. 19 to 00:00 UTC, Mar. 20) of the topographic conditions in East Asia.

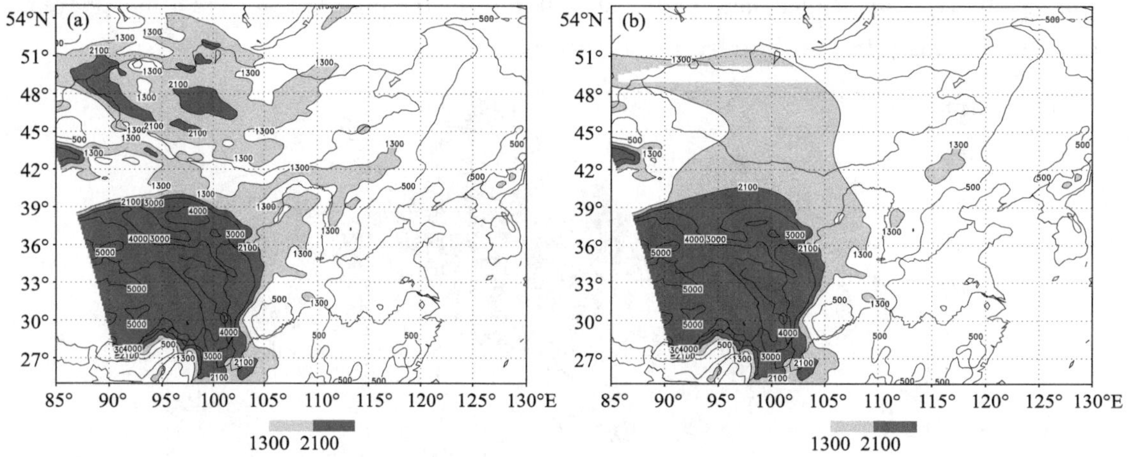

Fig. 4　The topography elevation in control run (a) and topographic sensitive test (b) (interval: 400 m)

5　Numerical simulation analysis

5.1　Dust weather simulation

Compared to the observed and simulated intensity of the dust storm (Fig. 5a-d), it can be identified that the outcomes of the control run preferably simulated the dust situations from 06:00 UTC, 12:00 UTC, 18:00 UTC of Mar. 19 to 00:00 UTC of Mar. 20, 2002. A large area along the boundary between Inner Mongolia and Mongolia can be seen in Fig. 5a which is covered by dense dust clouds. Meanwhile, comparing to the simulated output, a region with higher dust concentrations appears around the same area. At 12:00 UTC of Mar. 19, 2002, the suspended dust cloud was moving to eastward and becoming stronger than that of the original situations in Fig. 5b. Similarly, the higher value of the simulated dust concentration appears over the same region. From 18:00 UTC of Mar. 19 to 00:00 UTC of Mar. 20, 2002, the observed dust storm splits and moves in two directions, one heads to eastward, and then, reaches to south area of middle-eastern Inner Mongolia. The other one moves to the southern Shaanxi province. The same situation is simulated in control run (Fig. 5c-d). Therefore, the production and spread situation of the dust storm process is successfully simulated.

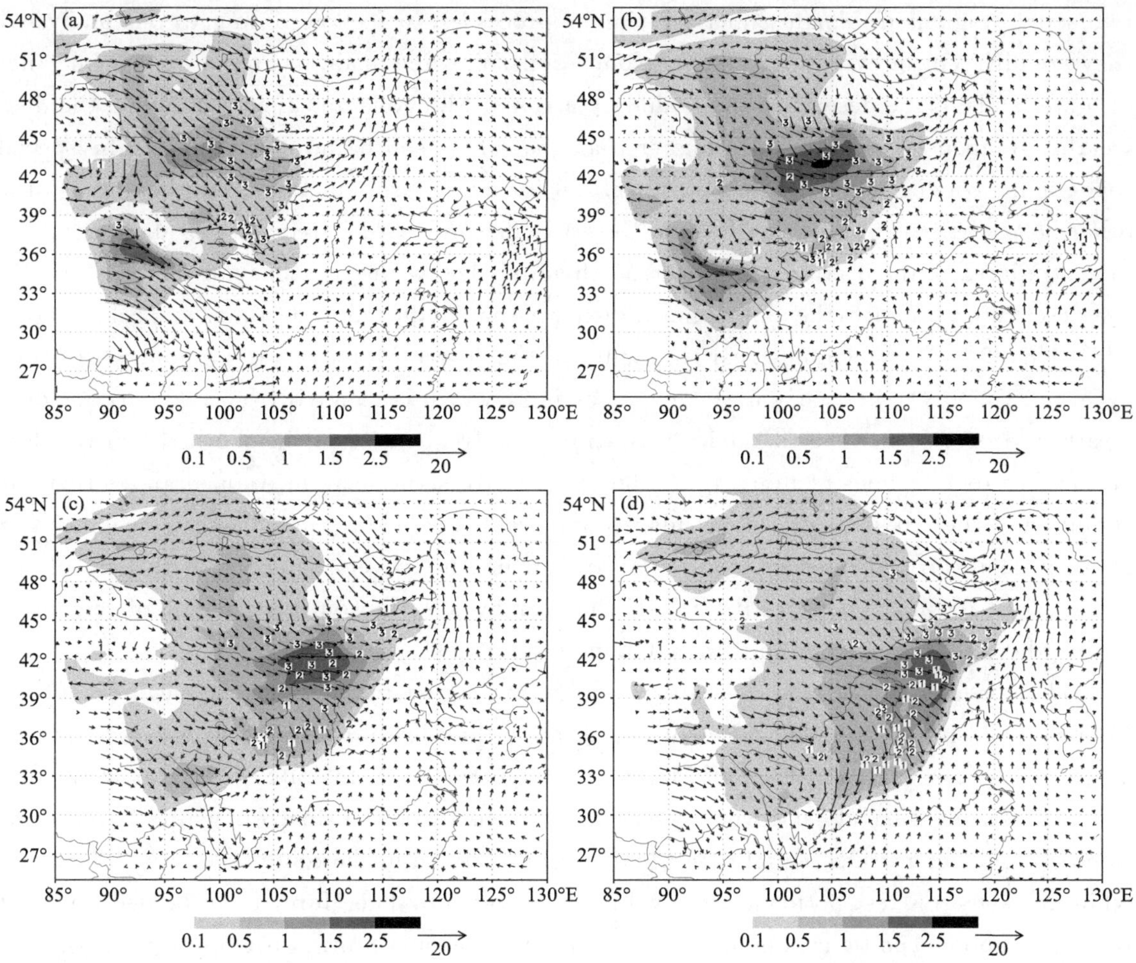

Fig. 5 The observed dusty weather, the simulated surface dust concentration and wind field at 10 m above surface in control run(Number 1, 2 and 3 represent floating dust, blowing dust and dust storm, separately; shaded area represents the unit of mg/m^3

a. at 06:00 UTC Mar. 19, 2002; b. 12:00 UTC Mar. 19, 2002; c. 18:00 UTC Mar. 19, 2002; d. 00:00 UTC Mar. 20, 2002)

5.2 Analysis on developing procedures of the dust storm event

Different characteristics of the dust event in dissimilar evolution periods are betrayed in Fig. 5a-d. From 06:00 UTC, Mar. 19 to 12:00 UTC, Mar. 19, 2002, the main feature of dust storm evolution is strengthening and the dust areas appeared within the region of southern Mongolia and western Inner Mongolia. On the next day, the main characteristics of the dust spread is the southeast moving activities from 18:00 UTC, Mar. 19 to 00:00 UTC, Mar. 20, 2002(Fig. 5 c-d). During this period, the dust storm intensity didn't change too much.

The evolution characteristics of the dust storm is a typical dust storm pattern observed in North China, which always starts from the Mongolia Plateau(the first dust source region), enhances itself quickly, and then spreads southeastwards and influences the downwind region. It cab be easily recognized by people that the developing patterns of dust storms of the two source regions in North China are similar processes. In fact, according to our analyses, they are different in evolutions and transportations. Viewing again in Fig. 1, we can find that in hinterlands of the deserts in the boundary areas of Mongolia and Inner Mongolia the dust weather frequency is significantly higher than that of its south and north edges. The special topographic conditions of the region are helpful for dust storm quickly strengthening. On the other hand, the situation in Taklimakan Desert is dissimilar. The frequency of dust weathers in central area of the Taklimakan Desert is obviously below the frequency of the south rim of the desert. For instance, the annual mean dust weather frequency is 33 times at Bachu station (78.34°E, 39.48°N) which lies near the center of the desert. Moreover, at Tazhong observatory (83.40°E, 39°N), the yearly mean frequency is 24 which is counted from the constant dust weather records from 1999 to 2002. On contrary, the frequencies of the south rim of the desert reaches to 100 times or more. It can be deduced from the facts mentioned above that the condition of the Taklimakan hinterlands is unfavorable for dust storm enhancement. In earlier stage of dust storm formed in the central area, it doesn't develop quickly. With help of the southward current, it enhances itself fast when it moves to the south rim region of the desert basin, where can provide an uplift dynamic power for dust storms.

The observed fact of the dust storm evolution process in the boundary region between Mongolia and Inner Mongolia are identified through the control run and sensitive test simulating the dust case at 12:00 UTC of March 19, 2002 (Fig. 6a-b). In spite of the fact which displayed the same dust cloud covered areas in the two tests, the dust intensities (concentrations) were different. The central concentration value revealed in the control run was 2.8 mg/m^3 while the sensitive test performed the value at 1.2 mg/m^3. Therefore, it can be deduced that the topographic conditions conduct a great influence on dust storm developments in the beginning period.

The topographic features can be seen in Fig. 2: in the hinterland of the Mongolia Plateau there are two quasi-east-west mountains. One is the Gobi Altay Mountains located in southern Mongolia while the other is the Yinshan Mountains in the central area of Inner Mongolia. The two mountains make a canyon between them. This canyon connects with another canyon lies at the south side of the Altay-Sayan Mountains. These two linked canyons stretch from west to east of the Mongolia Plateau with a length about 2000 km (90°—110°E). The low level currents over the canyon zone are always controlled by the narrow canyon conditions. Aimed in identifying this topographic influence, the simulating had been run with two different inputs of surface conditions, the control run and sensitive test (Fig. 7a-b). Compared to the two simulated outcomes, we can find that the current directions were quite dissimilar in the canyon region (rectangular area marked in Fig. 7). The wind directions in the first test displayed an obvious west-east style. Outside of the canyon, the winds headed to south unanimously both in the south and in the north areas (Fig. 7a). The situation in sensitive test was different. When the element of the surface conditions was not concerned in simulation, the wind directions inside and outside the canyon zone were performed mostly northwest-south-

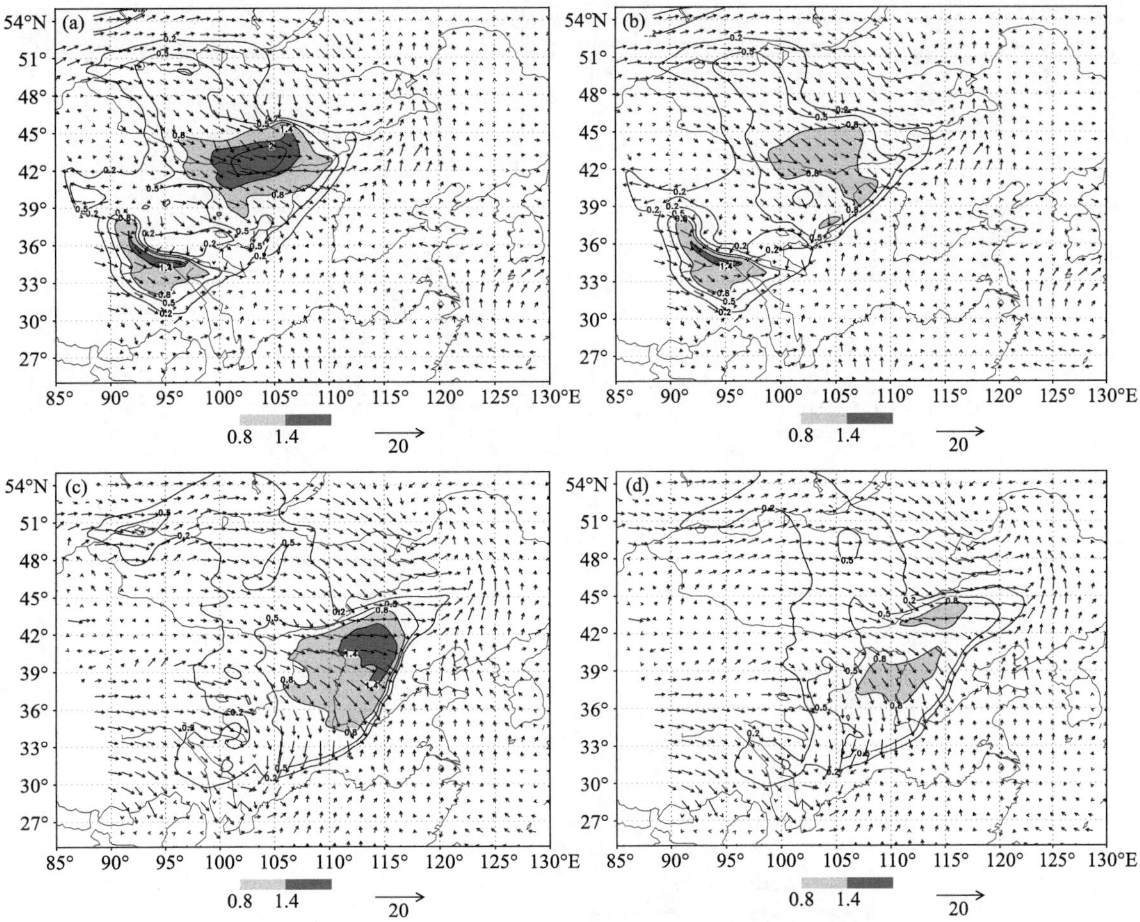

Fig. 6 Surface dust concentration (mg/m^3) and wind field at 10 m above the surface in control run at (a) 12:00 UTC Mar. 19, (c) 00:00 UTC Mar. 20 and sensitive test at (b) 12:00 UTC Mar. 19, (d) 00:00 UTC Mar. 20, 2002.

east style (Fig. 7b). That provides a powerful evidence for the conclusion of the topographic influence. Therefore, it can be recognized that the wind directions near the surface were changed by the topographic conditions of the canyon. It is most likely to produce a rounding effect of current in the canyon when the strong wind approaches to walls of the high mountains in both sides. Due to this rounding effect of current, the current is forced to become stronger, and then, a canyon low level jet (CLLJ) is formed. There is a fact can be proved by comparing the wind velocity changes in the two tests that wind speed in the canyon is affected significantly by the special topographic conditions. The difference between the simulated velocity in the canyon area of the control and the sensitive test is 2～4 m/s, the velocity in the first test is larger than that of the second test (Fig. 8). (Otherwise, dissimilar current situations can be viewed in Taklimakan Desert. Since the surface of the desert is relative even, especially in the central part of it, the topographic effect is small). Therefore, a fact can be identified from the comparing analysis that the canyon conditions of the region containing Mongolia and western Inner Mongolia is the crucial element for quickly wind strengthening.

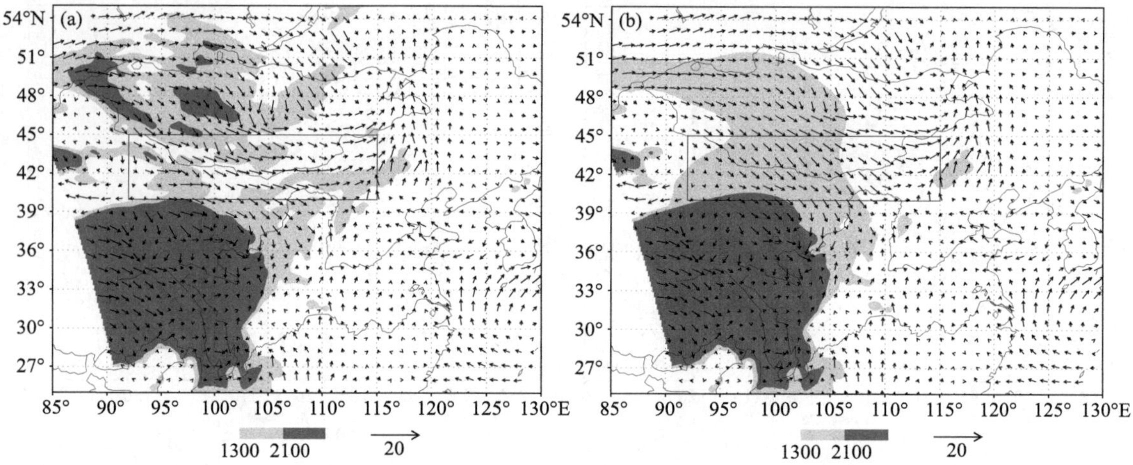

Fig. 7　The topographic elevation (shaded, above 1300 m) and the simulated surface wind fields of March 19, 2002 at 12:00 UTC (a: control run; b: sensitive run)

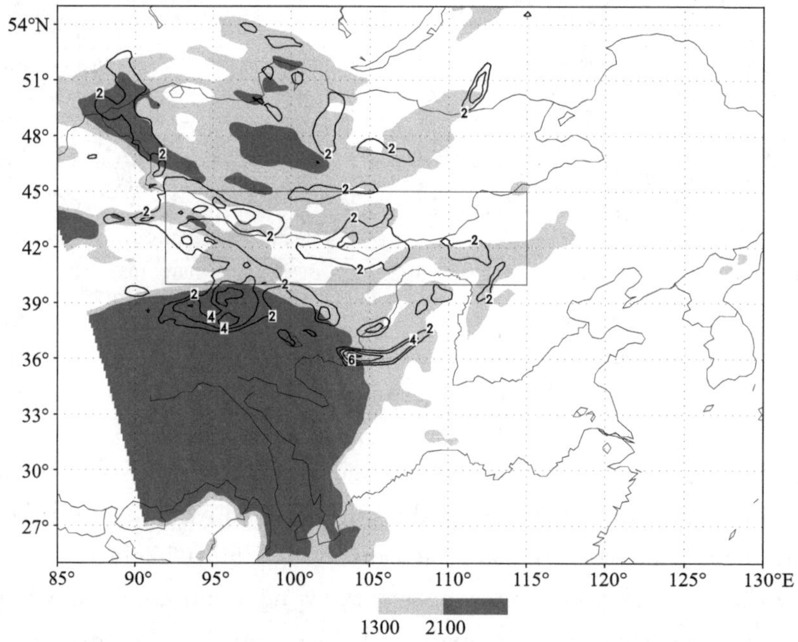

Fig. 8　The difference of wind speed (m/s) at 10 m above surface at 12:00 UTC Mar. 19, 2002 between control run and sensitive run. Shaded area is the topography elevation in control run (above 1300 m).

In Fig. 7a, a fact can be found that there is a great difference between the two currents directions. One is the obvious northwest down slope current over the south side of the Sayan Mountains, and the other is the CLLJ with actually west direction in the canyon located in the downwind region of the first current. Therefore, with the southeastern movement of the first current, it is blocked by the second one. Meanwhile, some dense dust clouds above the down slope area can be seen from the observed surface dust distributions of the period from 06:00 to 12:00 UTC on March 19, 2002 in Fig. 5. Under this condition, the northwest current, in fact, brought a lot amount of sand and dust heading to the southeast direction, and then, the sand and dust particles assembled around the canyon when the northwest current encounter the CLLJ. This dust assembled phenomenon also can be proved in the surface observations (Fig. 1). From the figure we can find a significant dust storm frequently attacked area which is fitted to the area around the canyon. Inevitability, the block effect of the canyon is another reason of the quickly strengthening of dust storm in the boundary region between Mongolia and Inner Mongolia, China.

Comparing with the orographic conditions of the Taklimakan Desert, there are many mountains and hills making an uneven surface in the area of South Mongolia and west Inner Mongolia (Fig. 2). The uneven surface condition induces orographic waves, which can produce ascending motion on the frontal slope of the mountains and descending motion at the back sides. On the frontal slope, the sand and dust can be easily lifted up into the upper air due to the ascending motion of the upward slope. That is one cause of why dust storms always are enhanced in the region of the frontal slope. More important, the orographic waves, in general, can expand upward to the free atmosphere and can uplift the dust particles into the upper air. This kind of impact, however, sparely can be seen in the Taklimakan Desert. Some obvious differences are revealed in the simulated outputs of the two tests at the zonal cross section of 42°N (Fig. 9a-b). The significant orographic waves can be seen in the control test. This wave effects not only appear in the boundary layer, but also expand into the free atmosphere. Furthermore, the friction area of the current and the earth's surface is surely enlarged by the uneven earth surface. That is also helpful for rising dust (Fig. 9a). Conversely, there is no short length waves displayed in the sensitive simulation (Fig. 9b).

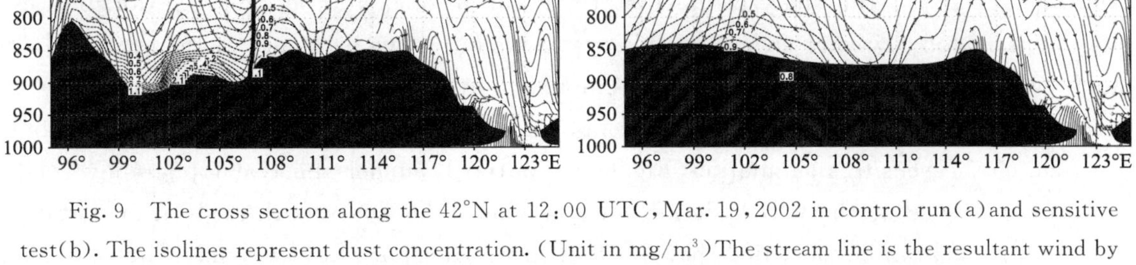

Fig. 9 The cross section along the 42°N at 12:00 UTC, Mar. 19, 2002 in control run (a) and sensitive test (b). The isolines represent dust concentration. (Unit in mg/m^3) The stream line is the resultant wind by u and $w*50$. The thick solid line represents the orographic wave induced ascending motion branches.

In ordinary recognition, it is possible that the variation of the simulated dust storm intensities in the two tests may be induced by the difference of cyclone intensity and the change of topographic conditions. In order to identify which one is the major influence factor, the simulated strength and evolution of the cyclone in the two tests are compared: at 12:00 UTC of March 19, 2002, the difference of cyclone central pressure values between the two tests is very small, 1006 hPa for the sensitive test and 1005 hPa for the control test. At 00:00 UTC on the next day, both of the central pressure values of those two tests are 997 hPa. Additionally, the central positions of the cyclone can be seen almost in the same point in model output wind fields of the two tests. Therefore, it can be confirmed that the variations of dust storm intensities between the outcomes of the two tests is mainly caused by the changes of topographic conditions.

The dust storms in the Mongolia Plateau are initially enhanced in the dust source region, and then, expanded to its downwind regions. This is absolutely dissimilar to the dust storm evolution pattern of the Taklimkan Desert, in where the dust particles are immediately transported to the southwards after they are lifted up into the sky. Only when the dusts encounter an obstacle, for example, the north slope of Altyn Mountains, it will become stronger. For the former one, the dust particles originally gather in the atmosphere over the region where can be recognized as a'dust accumulating container', the later transportation style of the suspended dust is crucially determined by the weather systems.

5.3 The spreading stage of dust storms

Generally, the dust storms spread continuously to the downwind regions after it had enhanced on the Mongolia Plateau. The impacts of the topography also played an important role in dust transportation in this stage. The obvious influence of the topography was revealed in the outputs of the two tests in which the surface dust concentrations are simulated at 00:00 UTC of March 20, 2002 (Fig. 6c-d). In the control run, the main body of the suspended dust moves directly to eastwards, otherwise, one more transporting branch obviously moving to southward can be viewed in the sensitive test. Therefore, a deduction can be got that the topographic conditions restricted the eastern movement of the dust storms occurred on the Mongolia Plateau.

The spreading direction of the dust clouds is crucially determined by weather system. The system of the analyzed dust storm event belongs to the type of Mongolia cyclone. The cyclonic circular air current always exists in the whole stages of the Mongolia cyclone evolution. There is a southwards component in the circular current which leads to southward transportation of the dust particles. The southwards component is revealed in output field of the sensitive simulation (Fig. 6d). Contrarily, under the real topography condition, the influence of the CLLJ significantly weakens the dust southwards transportation (Fig. 6c). It also can be identified that the southward transportation (v component) is weaker with the actual topography condition according to Fig. 10.

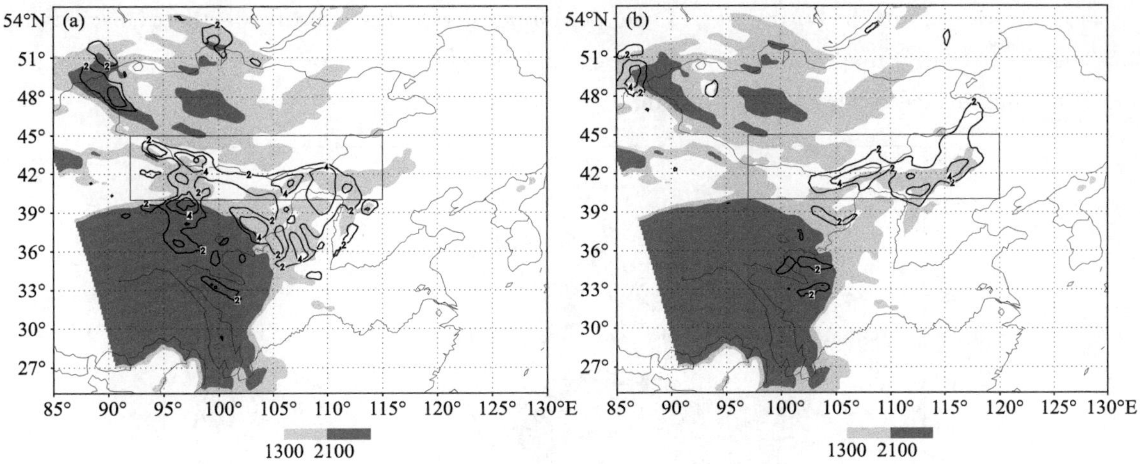

Fig. 10 The differences of u component(a) and v component(b) of wind fields(m/s) at 10 m above surface at 12:00 UTC of Mar. 19 2002 between control run and sensitive test. (the shaded area represents the topography)

The sub-frequent dust storm occurrence region is identified stretching from northeast edge of the Tibetan Plateau to the central part of the Loess Plateau(Fig. 1), where is not a dust source area. Evidently, most of the dust weathers observed in this region are caused by invading dusts sourcing from its upstream regions. There are two recognized topographic forcing elements which carry the dust particles moving southwards to the Loess Plateau, one is the down-slope air current in the south side of the Sayan Mountains, and the other is the rounding effect of the air current produced in the belt region around the north-eastern edges of the Tibetan Plateau. Due to the powerful impacts of the huge scales of the Plateau both in vertical and horizontal directions, the rounding current is very strong(Fig. 7). In the vertical direction, the current even can reach to the middle level of the troposphere. (The intensity of southing transportation in this region is stronger than that of the eastern movement of the dust clouds, because there is no effect of the canyon jet.) Otherwise, concerning only in the weather conditions, most of the dust eastern movements are led by the Mongolian cyclones while the southern transportations of dust particles are usually caused by cold fronts. The statistical number of the historical dust weather system investigation in North China shows that there are 76 percent of the severe dust storms are produced by the cold front. The Mongolian cyclone led dust storms is only 12 percent of the total number[10]. Putting all analyses mentioned above together, all factors which make great influence on the dust transportations in East Asia can be sorted into two kinds, one is impacted by weather systems consisting two weather types(the white arrows) and the other is influenced by topographic conditions including three terrain types(the black arrows). The dust transportation pattern induced by all five types is displayed in Fig. 11. Compared to all three topographic influence types, the rounding effect of current appears in the northeast edge of the Tibetan Plateau is the strongest one among them. In addition, the most frequent weather system causing dust storms in East

Asia is the cold front. In both cases of the rounding current and cold front, the dust particles are testified to be transported to south direction. Therefore, most of the transported dust deposit to the surface of the Loess Plateau.

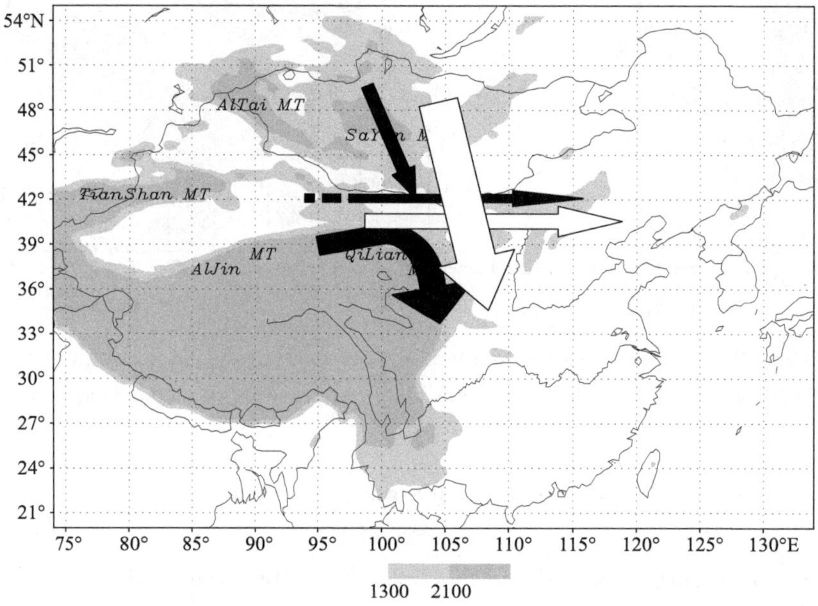

Fig. 11　The concept model of dust transportations in East Asia. (The black arrows represent the dust moving directions impacted by topography, and the white arrows indicate the transporting directions led by weather systems)

6　Summary

Based on investigation in the observation records of dust weathers in East Asia and by employing the method of numerical simulations, the influence of the topographic conditions on dust transportations in East Asia have been analyzed in this study. It has been shown that there are tow major dust storm source regions, one is in the boundary region between South Mongolia and the western part of Inner Mongolia, China and another is identified in the Tarim Basin. The most frequently dust storm happening area in the first one is confirmed in the central part of the region, and for the second one, is found in the south rim of the Taklimakan Desert. In addition, a sub-frequently dust storm attacked region is identified in the region which starts from the northeastern edge of the Tibetan Plateau to the Loess Plateau.

The major cause of dust quickly enhancement in the first source region is the combined effect of the favorable topographic conditions. The canyon feature of the region increases the wind speed, and then, forms a CLLJ from west to east direction. The amount of suspended dust particles, which are carried by the down slope current in the south side of the Sayan Mountains, are accumulated when they encounter the CLLJ in the canyon. Besides, the uneven earth surface condition induces orographic waves which are helpful for lifting more dust particles into the upper air. This process of dusts gathering in the atmosphere and then moving to the downwind regions can be recognized as a'dust accumulating container'.

The canyon condition of the south areas of Altay-Sayan Mountains is confirmed to be the crucial force for dust eastward transportations. In addition, the combined effect of the topographic force of the Tibetan Plateau and the down slope current in the south side of the Sayan Mountains is the major cause of the southward dust transportations. Otherwise, considering of the weather systems, the dust southward movement is mainly controlled by the cold fronts while the eastern transportation of the dust particles is led by the northwestern current behind the Mongolian cyclone during its fully developed stage. Therefore, the impacts of the rounding current existing in the northwest edge of the Tibetan Plateau combined with frequently cold front attacks bring most of the dust particles to southwards, which is the main reason of the transporting process are more likely gathering in the region which starts from the northern side of the Tibetan Plateau to the Loess Plateau.

References

[1] Shi G Y, and Zhao S X. Several Scientific Issues of Studies on the Dust Storms(in Chinese). *Chinese Jour. of Atmos. Sci.*, 2003, **27**(4): 591-606.

[2] Zhang X Y, Gong S L, Zhao T L, et al. Sources of Asian dust and role of climate change versus desertification in Asian dust emission. *Geophys. Res. Lett.*, 2003, **30**(24): 2272, doi: 10.1029 /2003GL018206.

[3] Zhang X Y, Wang Y G, Wang D, et. al. Characterization and sources of regional-scale transported carbonaceous and dust aerosols from different pathways in coastal and sandy land areas of China, *Geophys. Res.* 2005, **110**: D15301, doi: 10.1029/2004JD005457.

[4] Zhao T L, Gong S L, Zhang X Y, et al. A Simulated Climatology of Asian Dust Aerosol and Its Trans-Pacific Transport. Part I: Mean Climate and Validation. *J. Clima.*, 2006, **19**(1): 88-103.

[5] Hacker J P, Mckendry I G and Stull R B. Modeled downward transport of a passive tracer over western north America during an Asian dust event in April 1998. *J. App. Meteor.*, 2001, **40**(9): 1617-1628.

[6] Song M H, Qian Z A, Cai Y. Analysis of spring mean circulations for strong and weak dust-stormactivity years in China-Mongolia area. *Atca Meteor. Sinica*, 2007, **65**(1): 94-104.

[7] Jiang X G, Shen J G, Liu J T, et al. Observational and numerical simulation study of some weather factors leading a severe dust storm(in Chinese). *Acta Meteor. Sinica*, 2003, **61**(5): 606-620.

[8] Qian Z A, Cai Y, Liu J T. Some questionss in dust storm research in northern China. //The Workshop Corpus on Dust Storm Research(in Chinese): The Taiwan Mongolia-Tibet Committee Press, 2002, 63-76.

[9] Cheng C L, Wang Y C, Liu W D, et al. Study and case simulation of a regional dust model coupled with a nonhydrostatic dynamics model. *Chinese Science Bulletin*, 2004, **46**(19): 2007-2013.

[10] Liu J T, Qian Z A, Jiang X G, et al. A study on weather types of super severe dust storms in North China (in Chinese). *Plateau Meteorology*, 2004, **23**(8): 540-548.

[11] Anthes R A, and Warner T T. Development of hydrodynamics models for air pollution and mesometeorological studies. *Mon. Wea. Rev*, 1978, **106**(1): 1045-1078.

[12] Wang X L, Zhai P. The spatial and temporal variations of spring dust storms in China and its associations with surface wind and sea level pressures. *Atca Meteor. Sinica*, 2004, **62**(1): 96-103.

地面加热影响沙尘暴的数值模拟研究*

姜学恭[1] 李彰俊[2] 程丛兰[3] 胡英华[1] 张伯禹[1] 赵淑惠[4]

1. 内蒙古自治区气象台,呼和浩特 010051;2. 南京信息工程大学大气科学学院,南京 210044;
3. 北京城市气象科学研究所,北京 100875;内蒙古气象信息中心,呼和浩特 010051

摘　要:利用沙尘数值预报模式针对2002年3月18—20日典型沙尘暴过程中地面热通量影响沙尘暴的机制进行了数值模拟研究,结果发现:在荒漠化沙源地区,地面热通量以感热通量为主,它对沙尘暴的影响主要表现为导致沙尘暴产生明显的日变化。并通过加强地面风速、大气低层层结稳定度、行星边界层高度、地面摩擦速度的日变化发生作用。这一日变化强迫与蒙古气旋发展导致的沙尘暴持续增强的趋势共同作用使得沙尘暴逐渐增强的同时也具有一定的日变化波动。在时间和空间上,地面热通量的影响可以分为不同的阶段,在沙尘暴形成阶段主要体现在地面和行星边界层中;在沙尘暴扩展和传输阶段,地面热通量对上升运动强度的影响随时间推移从边界层中逐渐向上扩展到自由大气。这种直接影响的高度能够达到500 hPa左右。

关键词:沙尘暴,数值模拟,地面热通量

1　引言

沙尘暴作为一种复杂灾害性天气过程,其本质上是发生在大气边界层中,因此,对下垫面具有非常大的依赖性。这首先表现在荒漠化地表为沙尘暴提供了物质基础—沙尘源,地表的诸多物理特性,如土地利用、土壤类型、植被盖度、土壤湿度等都对下垫面起沙量具有重要影响,进而影响沙尘天气强度。其次,下垫面能够向大气的输送热量,为天气系统发生发展提供能量。不同于降水等其他天气过程,在沙尘暴过程中,沙尘暴的物质基础——沙尘粒子有赖于这一物质交换过程进入大气,因而地表与大气的能量交换过程是沙尘天气发生的关键环节。

目前,已有一些研究者针对下垫面物理特性如土壤类型、植被盖度、土壤湿度等与沙尘暴的关系开展了研究[1~3]。对于地面热通量与沙尘暴的关系,王劲松等利用NCEP再分析资料研究了我国典型沙尘暴过程中的地面感热加热、潜热加热分布特征,探讨了沙尘暴过程的热量的源汇特点[4~5]。孙军等则利用中尺度模式MM5模拟分析了沙尘暴过程中的地面热通量特点及其对沙尘暴的影响[6]。这些研究肯定了地面热通量对沙尘暴过程的重要影响。因此,本文拟利用沙尘数值预报模式,通过模拟分析和敏感性试验,结合沙尘暴过程个的沙尘粒子空间分布特点来研究地面热通量对沙尘暴的影响,并藉此揭示一些沙尘暴发生的规律。

2　天气过程概述

2002年3月18—22日沙尘暴过程是近些年影响我国的一次较为典型的强沙尘暴过

* 资助项目:国家自然科学基金(40565003)。

程[7]，影响地区不仅包括内蒙古西部荒漠化地区，还扩展至我国华北、东北及华东等地，另外，北京地区也出现了强沙尘暴。

本次过程是一次较为典型的蒙古气旋型沙尘暴过程。18 日 08 时(北京时间，下同)，在中亚地区对流层中层(500 hPa)有一低槽形成并逐渐东移加深；19 日 08 时槽线到达新疆东部(95°E)，在蒙古西部萨彦岭山地东侧形成地面蒙古气旋，同时扬沙和沙尘暴天气开始出现在这一地区；19 日 20 时，低槽东移并进一步加深，冷平流加大，蒙古气旋进一步加强，沙尘暴范围明显扩大并出现在我国内蒙古西部及我国西北地区。20 日，高空锋区增强的同时推进至河套一带，地面气旋随之东移并迅速增强，气旋冷锋自西北向东南扫过我国华北大部，导致这些地区强烈的沙尘天气；21 日，气旋东移，气旋冷锋扫过东北地区，沙尘暴推进至吉林、辽宁等地；之后，气旋减弱东移，影响我国大部地区的沙尘暴过程也逐渐结束。可见，本次沙尘暴过程历时长、影响范围广，而影响内蒙古及华北地区的时段主要是 3 月 19 日—21 日。

3 模式及模拟方案设计

研究采用北京城市气象科学研究所开发的耦合了天气模式和沙尘输送模式的沙尘天气预报模式[8]。其中，天气模式为中尺度模式 MM5V3，沙尘输送模式采用与 MM5 相同的 σ。

坐标，沙尘粒子的水平扩散、垂直扩散及差分方案与 MM5 相同。为了精确反映沙尘粒子的尺度特征，将沙尘粒径分为 16 档。模式的起沙过程利用参数化方法，起沙系数与土地利用、地表类型、植被盖度等因素有关。起沙通量由下式决定：$F=K\beta S\alpha\times U_*^4$，其中 F 为起沙通量，K 为比例系数(与沙尘谱宽有关)，β 为摩擦速度的非线性影响系数，S 为起沙系数，α 为地表湿度系数，U_* 为摩擦速度。起沙后，网格可分辨沙尘的上扬由上升运动和沙尘粒子的沉降速度决定。对次网格则由对流调整决定。沙尘沉降分为干沉降和湿沉降，大粒子的干沉降主要为重力沉降，小粒子沉降由湍流扩散和重力沉降共同决定。显式湿沉降由降水粒子对沙尘粒子的冲刷决定，隐式湿沉降由降水量决定。

模拟采用间隔 6 小时的 T213 分析和预报场作为初始场和侧边界，模拟区域中心经纬度 45°N 116°E，格距 30 km，格点数 131×121。模拟时段 3 月 19 日 08 时—3 月 21 日 08 时(北京时)。模式采用 Blackadar 高分辨率行星边界层参数化方案、云辐射方案和采用 5 层土壤模式。其中，控制试验中包括以上所有物理过程，去地面通量试验中将地面感热通量、地面地面潜热通量去掉，其他物理方案不变。

4 结果分析

4.1 对沙尘暴过程的模拟

图 1 为控制试验模拟的地面沙尘浓度随时间的变化，图中所标数值为实况沙尘天气等级(资料来自中国气象科学数据共享服务网的全球沙尘暴每日四次的沙尘天气观测资料)，其中 1、2、3 分别代表浮尘，扬沙和沙尘暴。图中，19 日 14 时，沙尘暴出现在蒙古国西南部及内蒙古西部，模拟的地面沙尘浓度稍弱于实况，但两者沙尘区范围差别不大。19 日 20 时，沙尘区向东扩展，强度明显增强，表现为沙尘暴站数明显增多，模拟地面沙尘浓度也明显增强，沙尘区范

围也与实况进一步吻合。20 日 02 时，沙尘暴扩展至内蒙古中部、宁夏及陕西、山西部分地区，模拟沙尘区与实况基本上一致。20 日 08 时，可以看到沙尘区扩展方向分裂为向东和向南，向东扩展至内蒙古东部偏南地区，向南扩展至山西、陕西大部地区，模拟的沙尘区与实况基本一致，只是向东扩展的沙尘强度稍弱于实况。总的来看，模式较为成功地模拟出了本次沙尘暴过程的发生、扩展及其强度变化。下文将基于控制试验和地面热通量敏感试验研究地面加热对沙尘暴的影响。

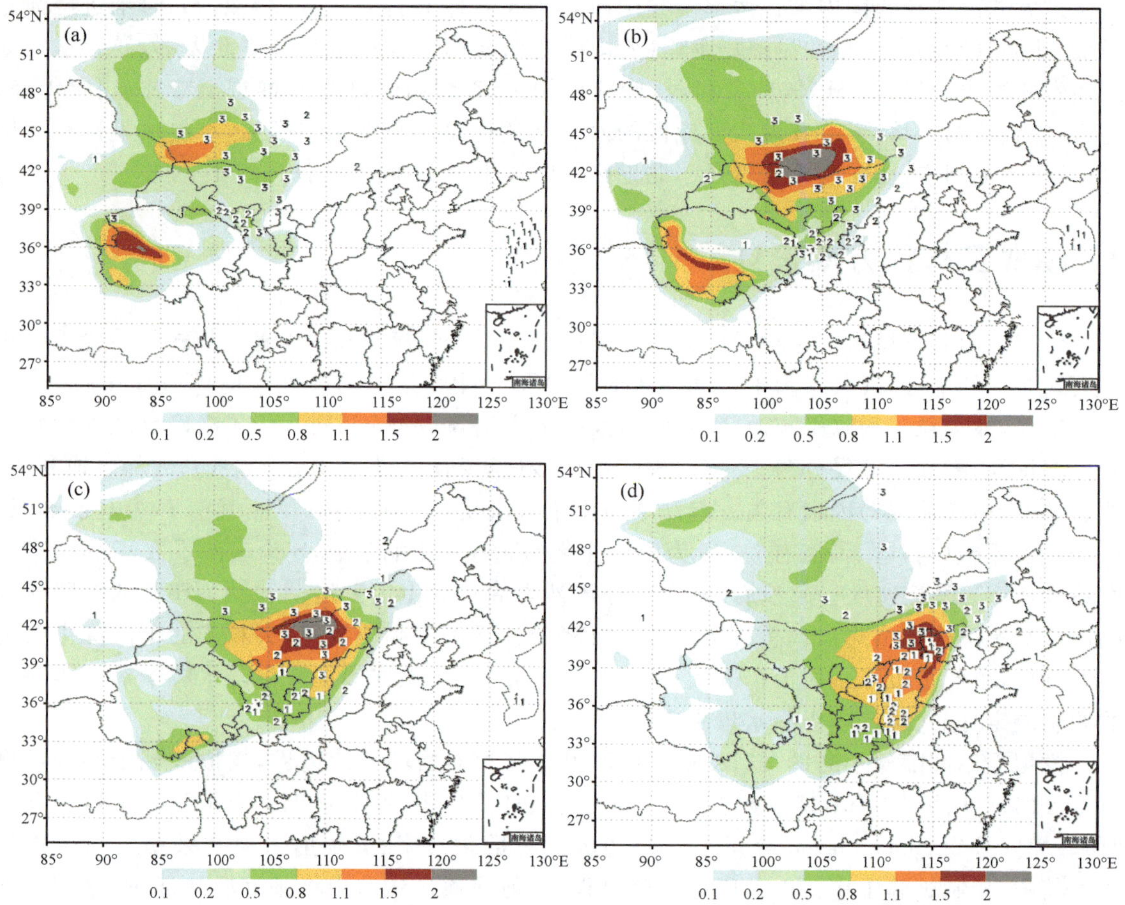

图 1　2002 年 3 月 19 日 14 时(a)、19 日 20 时(b)、20 日 02 时(c)、20 日 08 时(d)控制试验模拟的地面沙尘浓度(阴影区，单位 mg/m^3)及实况沙尘天气等级(1,2,3 分别为浮尘、扬沙和沙尘暴)

4.2　地面通量对沙尘暴强度的影响

从控制试验与去地面通量试验 19 日 14 时至 20 日 08 时地面沙尘浓度的强度变化可以发现，在 19 日 14 时(图略)，沙尘区主要位于蒙古西南部。此时是沙尘暴过程形成初期，控制试验沙尘浓度明显较强，中心值达到 1.1 mg/m^3 以上，同时沙尘区范围较大，而敏感试验沙尘浓度中心仅 0.2 mg/m^3 以上，且整个沙尘区范围明显小于控制试验。19 日 20 时(图 2a,b)，沙尘区向东偏南方向扩展至内蒙古中西部地区，控制试验沙尘浓度中心值达到 2.0 mg/m^3 以上，敏感试验达到 0.5 mg/m^3 以上，两试验差异进一步增强。20 日 02 时(图 2c,d)，沙尘区继续向东扩展，控制试验沙尘浓度中心值仍为 2.0 mg/m³ 以上，而敏感试验沙尘区在向东扩展的同时，沙尘浓度中心值反而增加至 0.8 mg/m³ 以上，这与控制试验形成明显差异。20 日 08 时，沙尘区向东扩展至内蒙古东部偏南地区，同时，沙尘区向南的扩展明显增强，到达山西、河北等地，但沙尘强度进一步减弱。控制试验沙尘中心浓度降为 1.5 mg/m³，而敏感试验沙尘浓度中心继续增强至 1.1 mg/m³。可见，地面热通量对沙尘暴强度具有非常大的影响，去掉地面热通量之后，不仅沙尘区范围有所减小，沙尘浓度也明显减小，两者沙尘浓度差异最大达到 5 倍之多。另外，在不同时间，地面热通量对沙尘暴的影响程度存在差异。

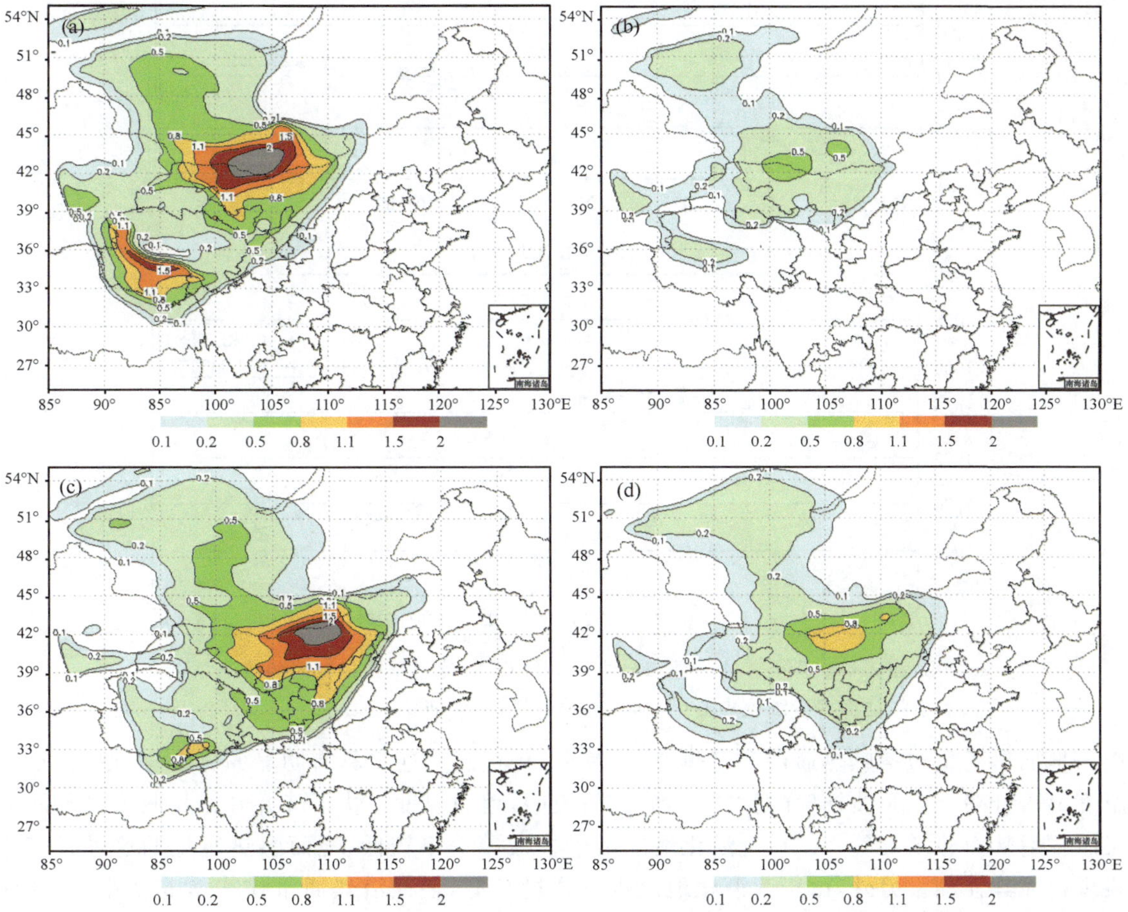

图 2　控制试验 2002 年 3 月 19 日 20 时(a)和 20 日 02 时(c)和去地面热通量试验模拟 19 日 20 时(b)和 20 日 02 时(d)的地面沙尘浓度(单位 mg/m³)

为了更清楚地表现沙尘强度的时间变化，图 3a 给出了 19 日 11 时至 20 日 08 时两试验地面沙尘浓度中心的时间变化。图中可以看到，两试验沙尘浓度中心最大差异是从 19 日 11 时—20 日 08 时去地面热通量试验沙尘浓度表现为单调增强，而控制试验则呈现单峰值的波动形态，傍晚前后为波峰。我们知道，对沙尘暴强度影响最大的因素无疑是天气系统，从本次过程蒙古气旋中心气压的变化可以看到(图 3b)，两试验气旋中心气压变化较为类似，除了 19 日 11—17 时略有增大外，基本上呈现单调减小。因此，在该时间段蒙古气旋的强度基本上是单调增强的。而这一变化是与去地面通量试验地面沙尘浓度中心的变化基本上一致的。可见，去地面通量试验沙尘浓度中心的变化主要是蒙古气旋强度变化的决定的。考虑到图 3 中控制试验沙尘浓度波动形态接近敏感试验的直线与单波峰简偕函数的叠加，因而地面热通量应该是沙尘浓度单波峰形态的形成原因，即导致沙尘暴强度出现明显的日变化。

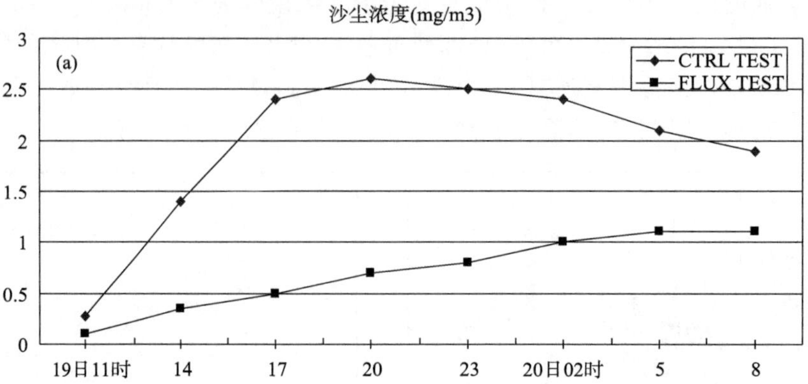

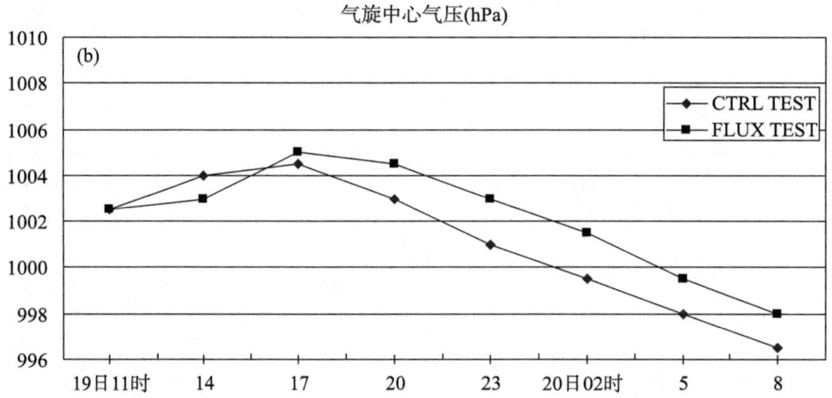

图 3　2002 年 3 月 19 日 11 时—20 日 08 时控制试验和去地面热通量试验模拟地面沙尘浓度中心(a)和气旋中心气压(b)演变

地面热通量包括感热通量和潜热通量，其中，地面感热通量具有明显的日变化：白天，地面温度显著升高并以长波辐射的方式加热大气；夜间，地面温度很快下降并开始接收大气释放的长波辐射从而冷却大气[9]。在荒漠化地区由于植被覆盖率较低，地面感热通量的这种日变化表现得尤其强烈。另外，由于荒漠化地区水汽含量较低，地面潜热通量往往是非常小的，地面热通量以感热通量为主。从控制试验和去地面通量试验 19 日 14 时(白天)和 20 日 02 时(夜间)地面热通量分布可以看到(图略)，19 日 14 时地面感热通量最大值达到 400 W/m^3 以上，且在蒙古南部、内蒙古中西部远大于其他地区；而在 20 日 02 时，上述地区地面感热通量变为负值(最小为－50 W/m^3)。而地面潜热通量基本在 100 W/m^3 以下，且日变化较小。可见，本次过程中，地面感热通量是地面热通量的主体，它不仅能够使沙尘暴强度增强，同时，其日变化也导致了沙尘暴强度产生相应的日变化。

4.3　地面通量对地面风速的影响

地面风速是影响沙尘暴强度的一个关键因素，而地面热通量对地面风速的影响能够间接影响沙尘暴。图 4 为 19 日 14 时和 20 日 02 时两试验地面 10 m 风场及全风速。图中，在发生沙尘暴的蒙古西部到内蒙古中西部一带，19 日 14 时控制试验地面风速(图 4a)明显强于敏感试验(图 4b)；而在 20 日 02 时，两试验地面风速强度基本上相当的。从图 5 两试验沙尘暴发生区地面最大风速时间演变也能够揭示这一特点。图中，19 日白天(14—17 时)控制试验地面最大风速较敏感试验强约 2.5 m/s，而在夜间(20—05 时)两试验地面最大风速基本上是相等的，之后两者差异又开始增大。即控制试验地面风速白天强、夜间弱的日变化明显强于敏感试验。可见。地面热通量对沙尘暴发生区域的地面风速也能够产生影响，且这一影响也主要表现为日变化。

地面热通量能够对地面风速产生影响的原因首先是地面加热能够使气旋冷锋前后温度梯度加大从而导致地面风速增强；另外，地面加热往往能够在干旱地区导致深厚混合层形成，而深厚的混合层有利于高空动量下传[10]，从而使地面风速增大。这一机制是荒漠化高原地区午后地面风速较大的主要原因。

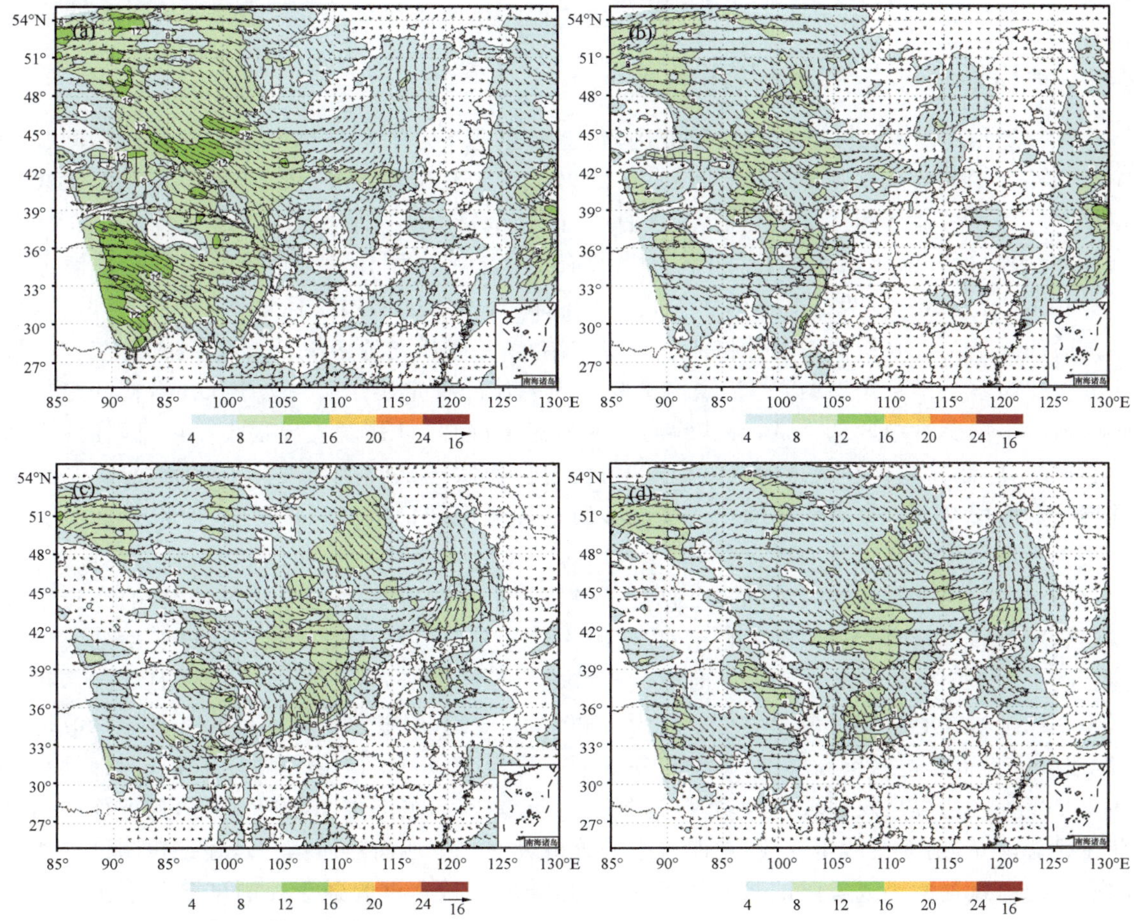

图 4　控制试验 2002 年 3 月 19 日 14 时(a)和 20 日 02 时(b)和去地面热通量试验 19 日 14 时(c)和 20 日 02 时(d)模拟的地面 10 m 风场及风速(单位 m/s)

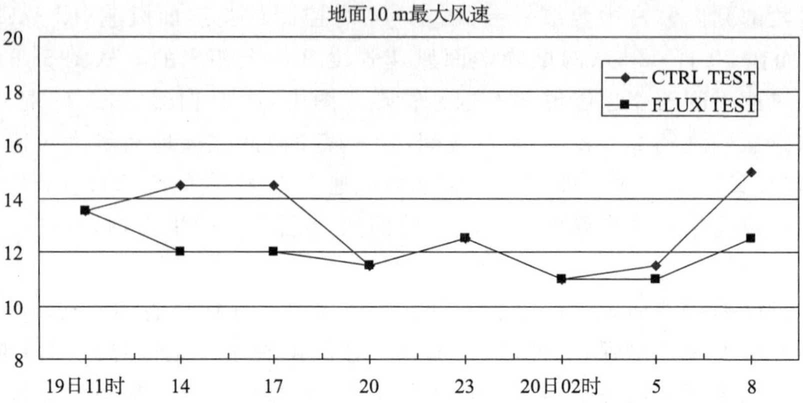

图 5　2002 年 3 月 19 日 11 时—20 日 08 时控制试验和去地面热通量试验模拟地面 10 m 风速极值(单位 hPa)

4.4　地面通量对大气层结的影响

大气层结是影响沙尘暴强度的重要因素,地面热通量对大气层结的影响同样能够间接影响沙尘暴。图 6 为两试验 19 日 14 时和 20 日 02 时 850 hPa 与 700 hPa 位温差(数值越大表示越稳定)。首先,通过图 2 可以知道在 19 日 14 时沙尘暴中心区域在基本上在蒙古南部,而 20 日 02 时则跨过我国北部边界位于内蒙古中西部。而图 6 中 19 日 14 时控制试验位温差为 −2～0 K,敏感试验则为 −8～−4 K。可见控制试验中该区域层结稳定度小于敏感试验(控制试验的层结为混合层而不是严格的不稳定层结,由于荒漠化地区缺乏水汽,地面加热往往形成混合层而不是不稳定层结[10],而敏感试验中为明显的稳定层结)。20 日 02 时,在沙尘暴中心区域——内蒙古中西部一带,两试验均表现为稳定层结,且强度差异很小。可见,在沙尘暴中心区域,控制试验存在明显的层结稳定度日变化,而敏感试验由于去掉了地面加热,基本上维持稳定的层结。这也表明地面热通量能够通过影响层结稳定度间接影响沙尘暴,而其影响方式也表现为加强沙尘暴的日变化。

与大气层结具有密切关系的一个物理量是行星边界层高度,而且与大气层结相比,行星边界层高度对沙尘暴具有更明确的指示意义,因为沙尘暴是一种主要发生在边界层中的天气过程。图 7 为 19 日 14 时和 20 日 02 时两试验行星边界层高度分布。图中,在 19 日 14 时,控制试验在沙尘暴中心区域行星边界层高度为 1000～1500 m,而敏感试验不足 50 m。到 20 日 02 时,控制试验行星边界层高度大为减小,基本上下降至 50 m 以下,值得注意的是敏感试验中,沙尘暴发生区边界层反而较白天增厚(达到 1000 m 左右),范围也明显扩大。这与前面的结果存在显著差异,前面对沙尘浓度、地面风速的分析表明去掉地面热通量只是导致控制试验沙尘浓度和地面风速减小或者两试验强度接近,而并未出现敏感试验强于控制试验的情况。而这一特点显然是由地面加热之外的其他因素导致的。究其原因,混合层平流可能是其一个重要形成原因,对比图 7b,d 可以发现,在没有地面热通量的情况下,在萨彦岭山地附近海拔较高的地区具有绝对高度较高的行星边界层,而内蒙古中西部地区此时并不具备形成深厚边界层的条件,20 日 02 时较深厚边界层的形成只能来源于系统发展或平流作用。由于行星边界层一般情况下表现为等熵的中性层结,因此,两者通常是一致的,而混合层空气在典型下坡条件下是可以被平流的,即上游高原地区绝对高度较高的混合层空气在沿下坡方向的低层气流的驱动下将被平流到下游的平原地区[10]。考虑到从萨彦岭山地沿西北—东南方向到内蒙古西部一带为明显的下坡地形，因此，这种情况下萨彦岭山地附近的绝对高度较高的混合层空气能够被平流到内蒙古中西部地区，即图 7d 中出现的情况。

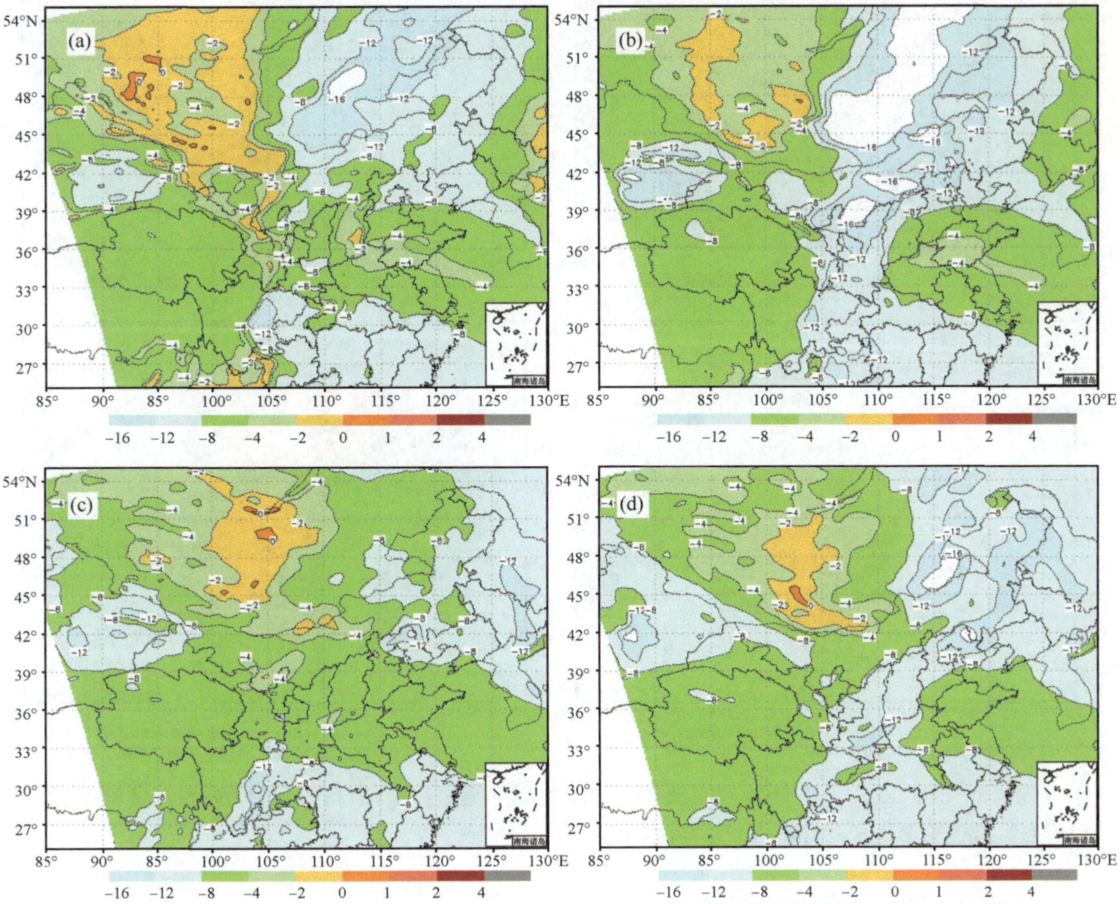

图 6 2002 年 3 月 19 日 14 时时控制试验(a)和地面热通量试验(b)和 20 日 02 时控制试验(c)和地面热通量试验(d)模拟 850～700 hPa 位温差值(单位 K)

混合层平流影响大气状态的核心部分是大气层结，对比图 6b，d 同样能够看出，在内蒙古中西部地区 19 日 14 时为较为稳定的层结，而 20 日 02 时稳定度显著减小，在没有地面热力作用的情况下，其只能来源于上游萨彦岭山地混合层空气的平流。同时这种平流作用也影响到沙尘暴的另外一个重要决定因素——摩擦速度。从两试验 19 日 14 时和 20 日 02 时地面摩擦速度分布可以看到(图略)，控制试验 19 日 14 时到 20 日 02 时地面摩擦速度最大值由 1.2 m/s 减小为 0.8 m/s；而敏感试验 19 日 14 时到 20 日 02 时地面摩擦速度最大值则由 0.6 m/s 增大为 0.8 m/s。可见，敏感试验中地面摩擦速度与行星边界层高度具有相似的变化。由于地面摩擦速度与大气层结的密切关系，因此，这一现象的产生原因也是容易理解的。可见，去掉地面热通量所导致的这些因素的变化都将间接地削弱沙尘暴强度的日变化。

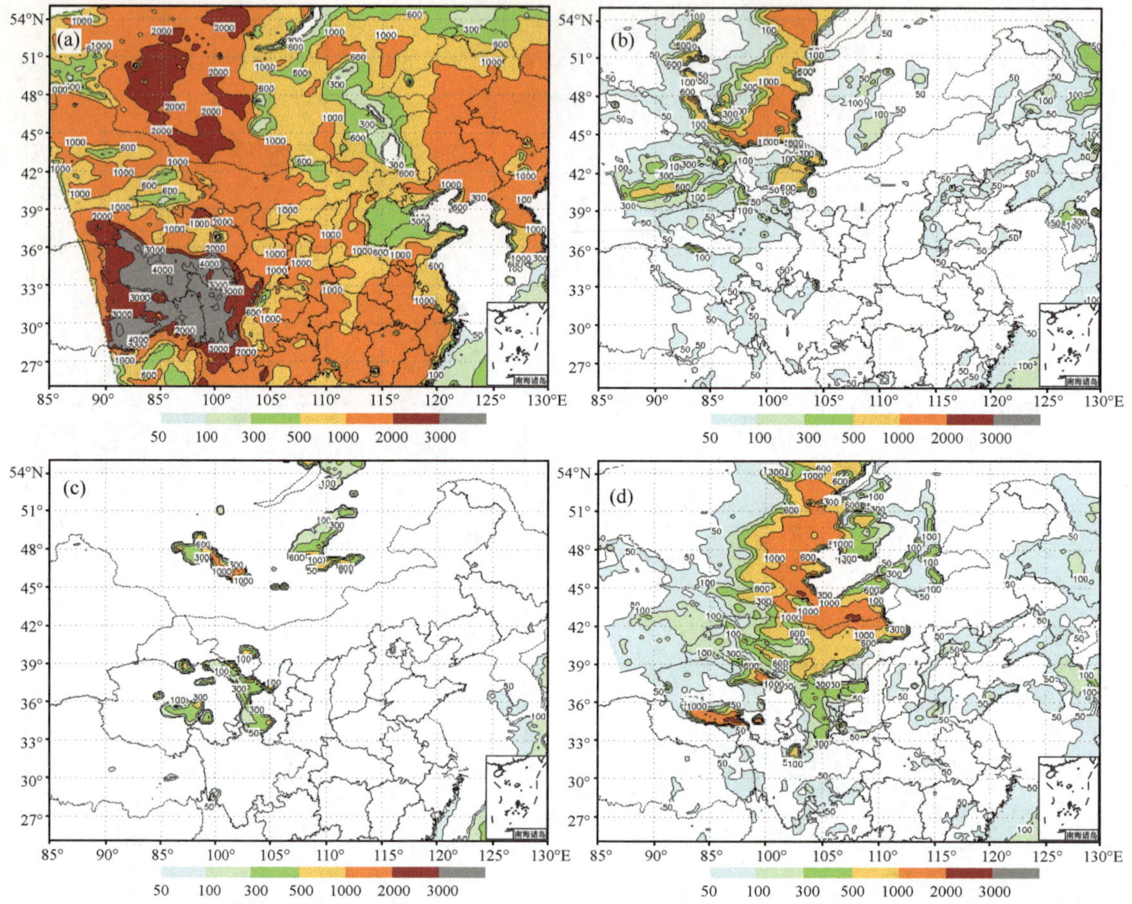

图 7　2002 年 3 月 19 日 14 时控制试验(a)和去地面热通量试验(b)和 20 日 02 时控制试验(c)和去地面热通量试验(d)模拟的行星边界层高度(单位 m)

4.5　地面通量对垂直速度的影响

上文分析的主要是地面热通量对地面及行星边界层中的一些天气要素的影响。沙尘在扬升至边界层之后,将进一步上扬到自由大气中,并通过传输、沉降等过程继续对下游地区产生影响。而这一过程中地面热通量的作用主要表现为对上升运动的影响。

图 8 为两试验 19 日 14 时至 20 日 08 时沿沙尘浓度中心(19 日 14 时为 43.5°N、19 日 20 时为 43°N、20 日 02 时为 42°N、20 日 08 时为 41.8°N)的纬向垂直剖面图。图中显示,在 19 日 14 时沙尘区主要位于沙尘源区(该阶段起沙较为旺盛),此时主体冷锋位于 113°E 附近,而副冷锋大约位于 108°E 附近。敏感试验中高空锋区前部没有明显的上升运动区,沙尘区主要位于高空锋区后部。而在控制试验中,沙尘区同样主要位于高空锋区后部,但在锋区前部 650～850 hPa 间有小范围的上升运动区,对应有小范围沙尘区存在,但与整个沙尘区相比所占比例很小。可见,这一阶段地面热通量对行星边界层上方的沙尘影响程度还比较小。(从两试验地面沙尘浓度的差异可以看出,此时地面热通量对地面沙尘暴强度的影响是比较明显的)。19 日 20 时,冷锋和副冷锋分别位于 114°E 和 108°E 附近,两试验沙尘区均已扩展至副冷锋前部,不同的是控制试验在副冷锋锋区前部 108°—113°E 范围内形成了从 925 hPa 贯穿至 650 hPa

的上升运动，而敏感试验副冷锋锋区前部的上升运动仅仅在 800 hPa 以下，且没有形成上下贯通，其垂直深度不足 100 hPa。对应的沙尘区在控制试验中不仅范围较大，同时已扩展至 600 hPa 以下，而在敏感试验中范围明显较小，仅达到 700 hPa 以下。至 20 日 02 时，两试验最明显的上升运动差异出现在冷锋前部 118°E 附近，控制试验上升运动从 700 hPa 以下一直达到 300 hPa 以下，形成了贯穿整个对流层中层的上升区，而敏感试验中上升运动并未出现这样的贯通，且强度偏弱。敏感试验中沙尘区未出现明显的向前或向上扩展。而控制试验中，沙尘区在对流层中下层明显向上和向前扩展，且已经达到接近 500 hPa。其形成原因显然是前一阶段沙尘区已经向上扩展至 600 hPa 附近，而之后形成的从 700 hPa 一直到接近 300 hPa 的贯通对流层中层的上升运动将导致沙尘继续上扬升(此时 500 hPa 以上有一沙尘区与其下方沙尘区相互割裂，应该来自其他沙尘源地)。20 日 08 时，控制试验中从 900 hPa 向上一直达到 400 hPa 的上升运动区依然维持，并表现为向前的倾斜上升区，而在敏感试验中上升区只出现在 900～750 hPa 和 600～450 hPa，且未实现上下连通；控制试验中沙尘区继续进行向前和向上扩展，并与 500 hPa 附近的沙尘区形成连通，而敏感试验沙尘区扩展高度反而进一步降低。

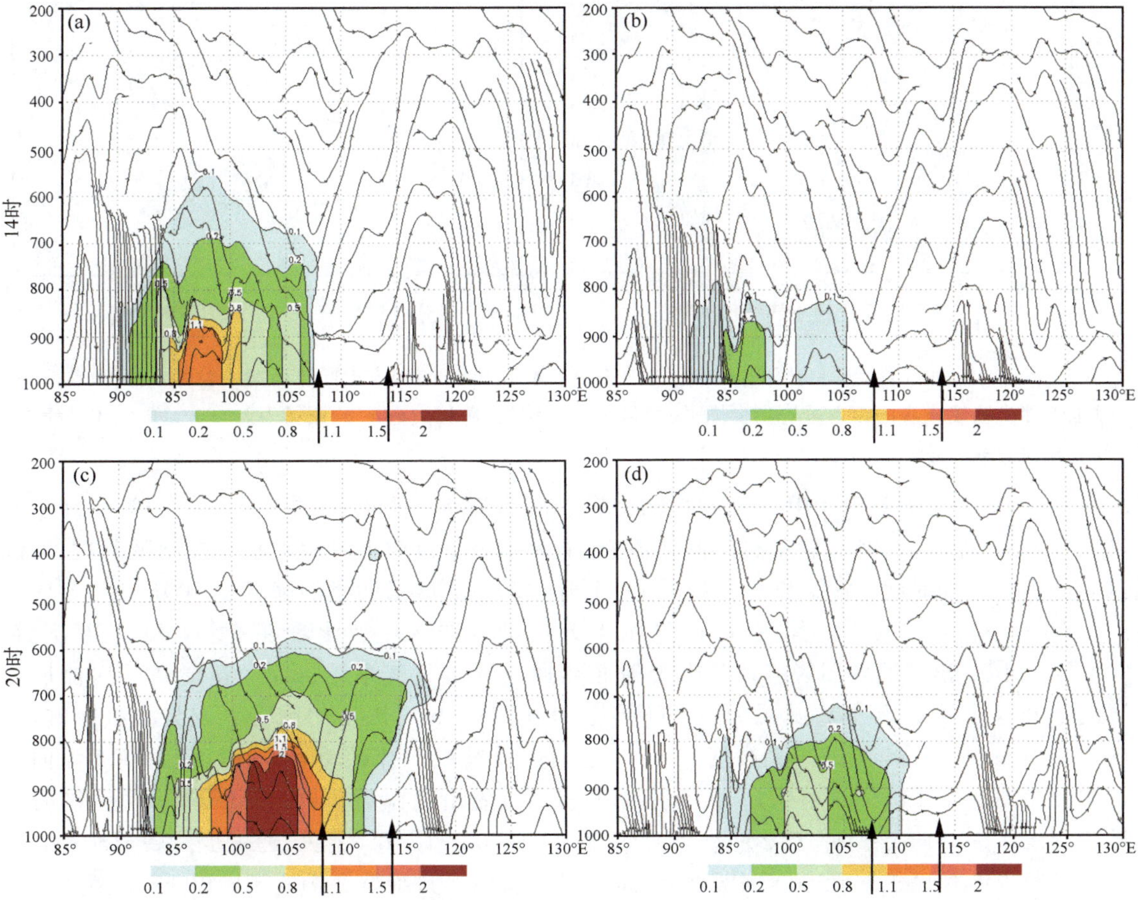

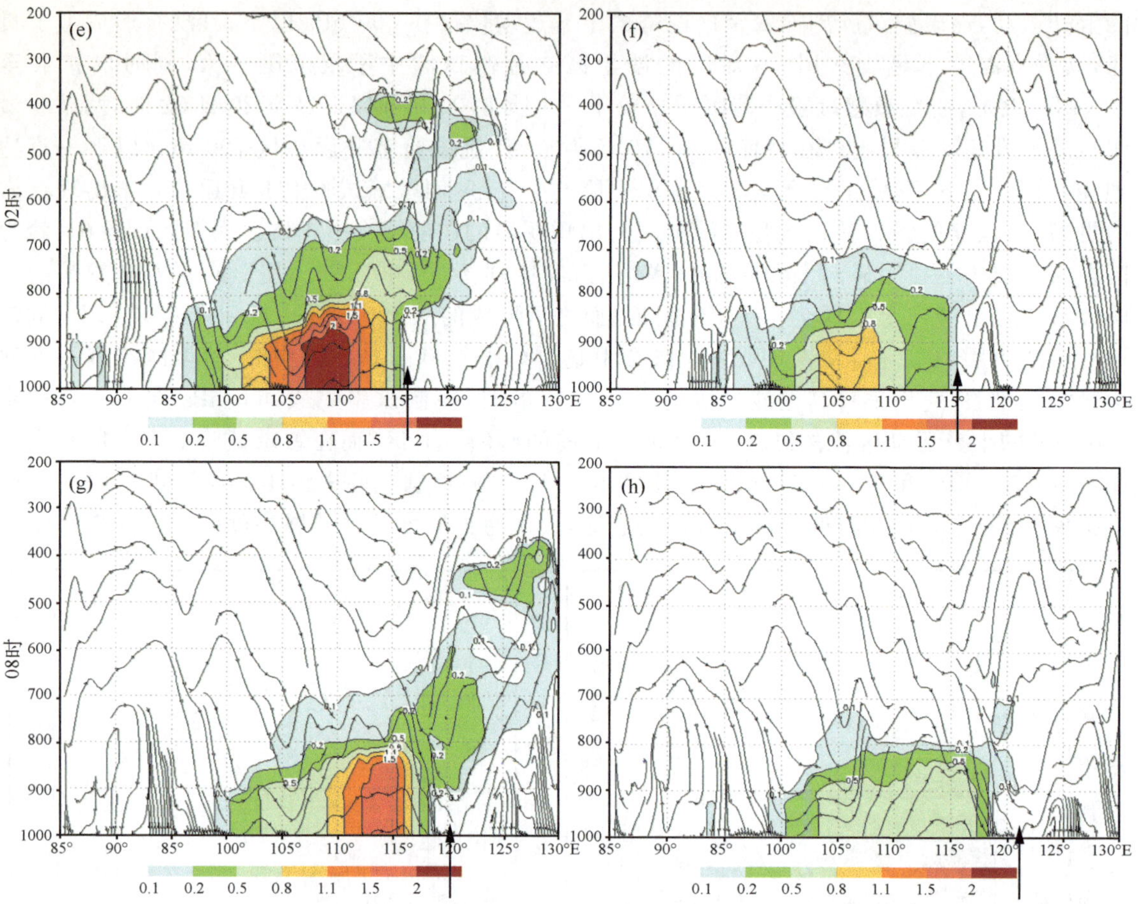

图 8　2002 年 3 月 19 日 14 时—20 日 08 时控制试验(a,c,e,g)和去地面热通量试验(b,d,f,h)模拟穿过地面沙尘浓度中心的纬向垂直剖面图,阴影区为沙尘浓度(单位 mg/m^3),流场为 $u,w*100$ 合成。箭头代表地面冷锋位置。

可见,地面热通量对上升运动的影响首先表现为影响上升运动的强度。图 9 为强沙尘暴发生区敏感试验与控制试验垂直速度差值(负值表示敏感试验上升运动小于控制试验)及控制试验沙尘浓度的时间演变。图中可以看到,上升运动减弱区最明显的高度在 700～600 hPa 之间,减小达 8 cm/s 以上。其次,图 9 显示上升运动减弱区主要出现在 500 hPa 以下,而在 300～400 hPa 出现的减弱区应该是由于地面热通量影响天气系统进而间接导致的对流层高层的上升运动改变,不妨将其视为一种间接影响,而 500 hPa 以下的则可界定为直接影响,因此,500 hPa 为这一影响的最大高度。另外,地面热通量对上升运动的影响是随时间逐渐向上扩展的。图 9 中 19 日 18 时前后,上升运动的减弱首先出现在 750～900 hPa 间,之后,随时间推移减弱区逐渐向上扩展,至 20 日 01 时到达 600～700 hPa 间,之后,进一步扩展至500 hPa。控制试验中沙尘区扩展到 500 hPa 以上,而敏感试验始终只维持在 700 hPa 以下,地面热通量对上升运动的影响显然是其产生的主要原因。

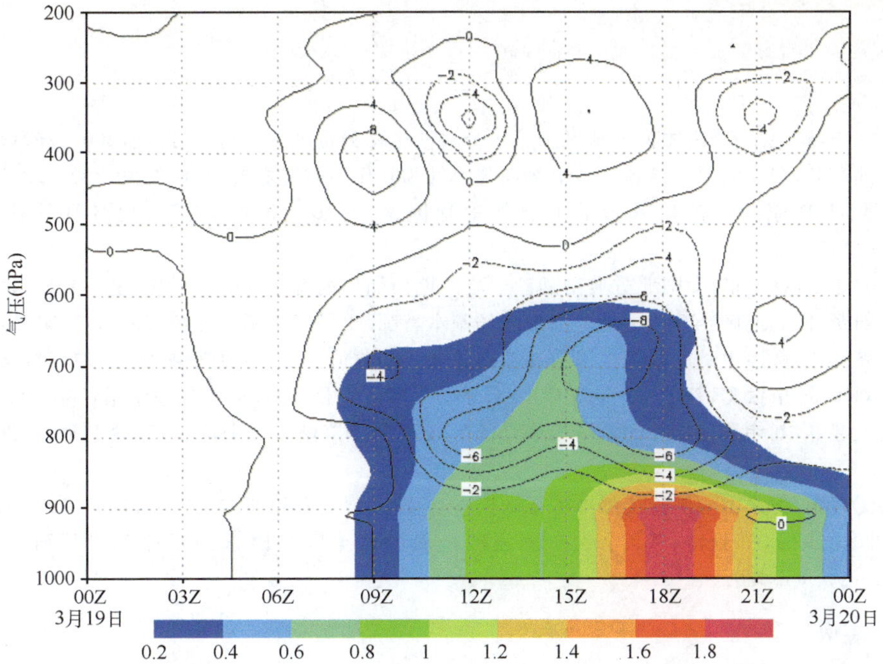

图 9 强沙尘暴区(110°E、43°N)2002 年 3 月 19 日 08 时—20 日 08 时去地面热通量试验与控制试验及垂直速度差值(FLUX-CTRL,等值线,单位 cm/s)及控制试验模拟的沙尘浓度(阴影区,单位 mg/m³)时间—高度剖面图

5 小结

本文利用沙尘数值预报模式并针对 2002 年 3 月 18—20 日影响我国的强沙尘暴天气过程,对地面热通量影响沙尘暴的方式、程度进行了敏感性试验研究,结果表明:

(1)在蒙古西部、我国内蒙古中西部等沙源地区,地面热通量以感热为主。地面热通量对沙尘暴的影响表现为两种方式,首先是导致沙尘暴增强,更为重要的是使沙尘暴产生明显的日变化。地面热通量能够导致地面风速、大气低层层结稳定度、行星边界层高度、地面摩擦速度等产生明显的日变化进而加强沙尘暴的日变化。

(2)在沙尘暴过程中存在两种主要的强迫因素影响沙尘暴的强度和演变,即天气系统因素和地面热通量因素。在本次过程中,蒙古气旋的持续增强使沙尘暴具有一种持续增强的趋势,它与地面热通量导致的沙尘暴强度日变化趋势共同作用使得沙尘暴强度一方面逐渐增强,同时具有一定的日变化波动。另外,混合层平流机制导致的低稳定度、深厚行星边界层、强地面摩擦速度在强沙尘暴区的维持也是沙尘暴持续增强的原因之一。

(3)地面热通量对沙尘暴的影响可以分为不同的阶段,上述的地面热通量影响主要发生在沙尘暴形成阶段,并主要集中在地面和行星边界层中。在沙尘暴的进一步扩展和传输阶段,地面热通量能够对上升运动强度产生影响且这种影响随着时间推移从边界层中逐渐向上扩展到自由大气。这种直接影响的高度能够达到 500 hPa 左右。

参考文献

[1] 李彰俊，李宁，顾卫．内蒙古中西部地区土壤水分对沙尘暴的贡献．地球科学进展，2005，**20**(1)：24-28.
[2] 胡樵隐，孙菽芬，郑元润，等．稀疏植被下垫面与大气相互作用研究进展．高原气象．2004，**23**(3)：281-296.
[3] 许炯心．黄土高原地区沙尘暴形成的自然地理因素：Ⅱ多元回归分析与地带性特征．中国沙漠，2005，**25**(4)：552-556.
[4] 王劲松，俞亚勋，赵建华．中国北方典型强沙尘暴的地面加热场特征分析．中国沙漠，2004，**24**(5)：599-602.
[5] 王劲松，刘贤．西北地区春季沙尘暴地面加热场基本特征．干旱区资源与环境，2003，**17**(5)：1：6.
[6] 孙军，姚秀萍．一次沙尘暴过程锋生函数和地表热通量的数值诊断．高原气象，2002，**21**(5)：488-494.
[7] 肖贤俊，刘环珠，宋振鑫，等．2002年3月19日沙尘暴爆发条件分析．应用气象学报，2004，**15**(1)：1-9.
[8] 程丛兰，王迎春，刘伟东，等．与动力场相耦合的区域沙尘分档模式及个例模拟．科学通报，2004，**46**(19)：2007-201.
[9] 姜学恭，沈建国．内蒙古两例持续型沙尘暴的天气特征．气候与环境研究，2006，**11**(6)：702-712.
[10] 姜学恭，沈建国，刘景涛，等．导致一例强沙尘暴的若干天气因素的观测和模拟研究．气象学报，2003，**61**(5)：606-620.

A Numerical Study on the Influence of the Surface Heat Flux on Dust Storm

JIANG Xuegong[1]　LI Zhangjun[2]　CHENG Conglan[3]
HU Yinghua[1]　ZHANG Boyu[1]　ZHAO Shului[4]

1. Inner Mongolia Autonomous Region Meteorological Observatary，Huhhot China 010051；
2. Inner Mongolia Autonomous Regional Meteorological Bureau，Huhhot，China 010051；
3. Institute of Urban Meteorology，CMA，Beijing 100089；
4. Inner Mongolia Autonomous Region Meteorological Informational Center

Abstract　Based on a numerical study on the dust storm process occurred from Mar. 18 to Mar. 20, 2002, the influence of the surface heat flux on the dust storm was conducted. The result shown: In the desert region the surface sensible heat flux is important and it enhances the diurnal change of the surface wind speed, the atmospheric stability at lower troposphere, the height of Planet Boundary Layer(PBL) and the surface friction velocity in the areas where the severe dust storm occurs and so enhances the diurnal change of dust storm. It works coupling with the effect, which is conducted by the Mongolia cyclogenesis and will induces a continual strengthening of dust storm, so that the dust storm exhibits not only a certain extend diurnal change but also a gradually strengthening process. On the temporal and spatial aspects, the influence of surface heat flux can be separated into two phases. In the dust storm formation phase, the influence of surface heat flux focuses on the PBL. When the dust storm expands towards the surface heat flux enhances the ascending motion from the lower troposphere up to the middle troposphere and so strengthens the dust particles rising and transportation. This direct influence that the surface heat flux conducts on the ascending motion can touches a height of 500 hPa.

Key words: Dust storm, Surface heat flux, Numerical simulation

地形影响蒙古气旋发展的观测和模拟研究*

姜学恭[1] 沈建国[2] 刘景涛[1] 陈受钧[3] 王迎春[4]

1. 内蒙古自治区气象台,呼和浩特 010051;2. 内蒙古自治区气象局 010051);
3. 北京大学物理学院大气科学系,北京 100871;4. 北京城市气象科学研究所,北京 100089

摘 要:本文针对影响蒙古气旋发展的地形因素进行了观测和模拟研究,结果表明:地形影响使蒙古气旋可以依据冷空气对地面气旋冷锋的强迫分为触发阶段和发展阶段。在气旋的发展阶段,斜压不稳定是其主要强迫机制。伴随地形对低层冷空气阻滞程度的变化蒙古气旋经历了从缓慢发展到剧烈加强的过程。阿尔泰—萨彦岭山地通过对低层冷空气的阻滞使山地上空等熵面更为陡立,加强了对流层低层的斜压强迫,从而使斜压强迫的涡度增长的向低层聚集,导致气旋发展强度增强。阿尔泰—萨彦岭山地及其南侧形成的峡谷地形对低空急流的位置、范围、强度及演变过程具有较重要影响。本次过程对流层高层位涡平流也是气旋发展的一个强迫因素,但其影响较小。

关键词:蒙古气旋,数值模拟,地形影响

1 引言

蒙古气旋是冬、春季对我国北方具有重要影响的天气系统,往往造成大风、大雪、寒潮、沙尘暴等强灾害性天气。其中,蒙古气旋是引发沙尘暴天气的重要系统,如在2001年冬春季发生的21次沙尘暴天气过程中,绝大部分(18次)是由蒙古气旋的强烈发展所致。

蒙古气旋是斜压性很强的极锋上的波动,同时,它又是较典型的背风坡气旋。陈受钧等[1,2]的研究中曾将蒙古气旋分为地形诱发(Terrain Induced)和地形影响(Terrain Modified)的蒙古气旋,前者为低层斜压强迫起主导作用;后者高层位涡平流起主要作用,强度较强,也较为深厚。Buzzi[3]曾经将阿尔卑斯背风坡气旋的发展阶段分为1)触发阶段:从地形对低层冷锋形成阻塞至冷锋越过山地形成(地形诱发的)有效位能的突然释放(e折时间尺度为6小时);2)缓慢增长阶段:即斜压发展阶段(e折时间尺度为25~30小时)。Zupanski等[4]则将背风坡气旋的发展分为三个阶段,即:1)触发阶段,高空急流是主要的强迫因素;2)发展中期:发展强度与低层锋生密切相关,显示了斜压强迫的重要性;3)生命后期:与高层的急流和位涡密切联系。Mattcks等[5,6]总结得出所谓"不同阻塞 Differential Block"理论解释背风坡气旋的触发。当高低层均有扰动形成时,高空急流出口区左侧的辐散区将强迫产生地面减压,这一地面减压往往由于低层的冷平流而被抵销,不会造成明显的地面减压。当低层冷空气遇到山地被地形阻塞(Block)时,高层系统继续向前移动,原有的耦合被打破,由于失去低层冷平流的制约,高空急流出口区左侧的辐散强迫将迅速导致山

* 资助课题:中国科技部重点科技攻关项目(2000-k01-03-02)和科技部公益类项目2001DIB20103、北京市科委项目(955300600)。

地背风坡大气质量减小，形成强上升运动、低层辐合和正涡度产生，诱发背风坡气旋。其时间尺度为冷锋形成阻塞到冷空气最终越过山地这一时段，一般为 6～12 小时。可见，背风坡气旋的形成综合了地形影响、斜压作用、高层强迫等因素，而在触发阶段往往存在高低层系统的相互作用。

本文针对一例蒙古气旋强烈发展过程，利用常规观测资料和数值模拟方法，主要探讨了地形对蒙古气旋发展产生的影响，希望揭示导致蒙古气旋发生的一些规律。

2 过程概述

本次过程蒙古气旋首先生成于蒙西山地西侧，之后东移上山地减弱，到达背风坡后又得到发展加强。5 日 02 时(北京时，下同)有气旋生成于阿尔泰—萨彦岭山地西北侧(中心位于 60°N，80°E，989 hPa)。5 日 08 时，高空槽移上山地，地形影响使高空槽减弱变平；地面气旋分裂为多个中低压散布于山地。至 5 日 20 时，高空槽继续减弱并表现为短波槽形态，对流层低层温度槽明显落后于高度槽，与之呈反位相分布(图 1)；地面有弱的中低压位于山地(48°N，94°E 和 48°N，98°E，992 hPa)。6 日 02 时，地面中低压在山地东部合并，形成锋面气旋(995 hPa)。6 日 08 时，高空槽下坡加深，地面气旋到达背风坡并开始发展加强(48°N，104°E，991 hPa)。6 日 20 时，对流层低层槽线已移过山地，移速减慢，高度槽加深形成切断低涡。温度槽与高度槽位相差减小，表现出典型的斜压结构，有强冷空气平流到高度槽中；地面气旋已加强至 984 hPa。6 日 20 时—7 日 02 时，气旋移动趋于停滞，强度快速增强(7 日 02 时为 978 hPa)。7 日 08 时后，500 hPa 形成低涡，斜压性消失，气旋锢囚(977 hPa)。气旋随之东移减弱并逐渐消失(7 日 20 时为 992 hPa)。在本次过程中，气旋强烈发展时段 6 日 08 时—7 日 02 时，中心气压加深 13 hPa/18h(991～978 hPa)。尽管未达到爆发性气旋的标准，但就陆地气旋而言，已是较罕见的发展速度。

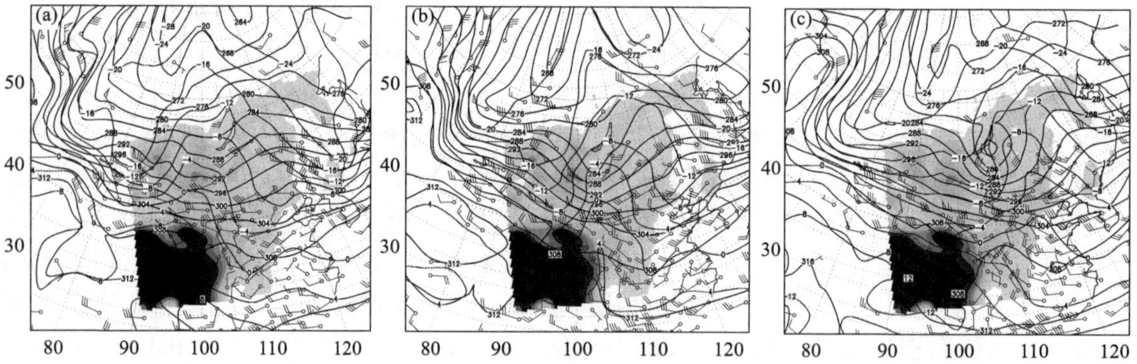

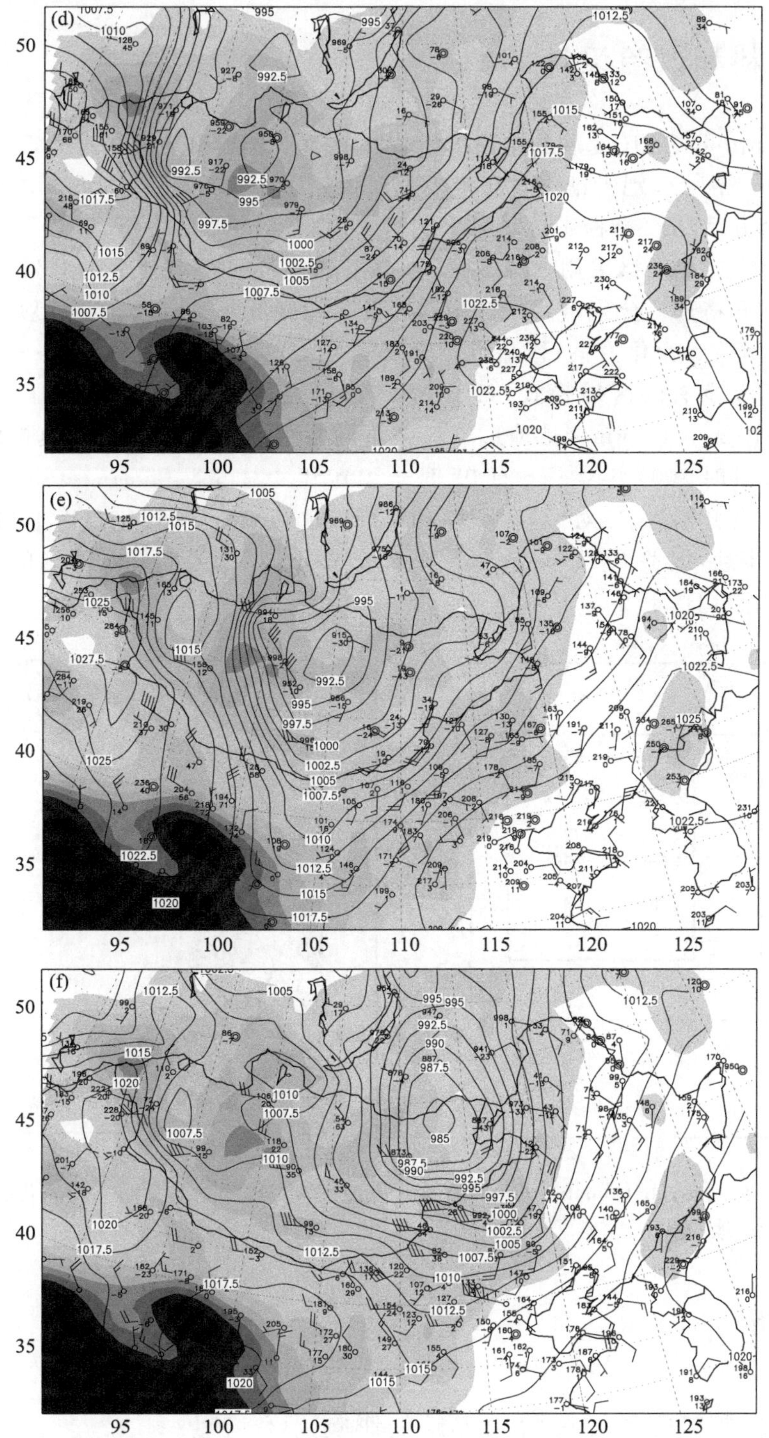

图 1 700 hPa 实况天气图(a)2001 年 4 月 5 日 20 时;(b)2001 年 4 月 6 日 08 时;(c)2001 年 4 月 6 日 20 时;和地面天气图(d)2001 年 4 月 5 日 20 时;(e)2001 年 4 月 6 日 08 时;(f)2001 年 4 月 6 日 20 时。图中 700 hPa 实线为高度。地面图上标注为气压和 3 小时变压。阴影区为地形高度

3　模式、资料和试验设计

采用 Pan State/NCAR 发展的中尺度模式非静力 MM5V3[7]。模式的垂直坐标 $\sigma=(p-p_t)/(p_s-p_t)$，p_s 和 p_t 分别为地面气压和模式层顶气压($p_t=50$ hPa)，垂直方向为 24σ 层。采用三重双向嵌套方案(图 2a，模拟区域)，大网格采用 90 km 水平格距，中心为 45°N，110°E，模拟范围 45×41 格点；中网格采用 30 km 格距，模拟范围 73×73 格点；细网格采用 10 km 格距，模拟范围 139×76 格点。分别覆盖气旋中心区及其南侧低空急流区。湿过程采用显式水汽方案和 Grell 积云对流参数化方案；行星边界层采用 MRF 高分辨参数化方案；采用云辐射方案。

初始场采用北京有限区域模式(HLAFS)客观分析资料插值得到。地形资料分别采用 NCAR 30 min、10 min、5 min 全球地形资料按不同格距插值生成(图 2b 模拟区域地形)。模拟积分时段为北京时 2001 年 4 月 6 日 08 时—2001 年 4 月 8 日 08 时。

模拟试验方案分别为(1)对照模拟(Control Experiment，CE)：包含以上物理方案和参数选择。(2)平面地形试验(Plane Terrain Experiment，PTE)：32°N 以北的地形高度取为 500 m，而其他参数选择不变。

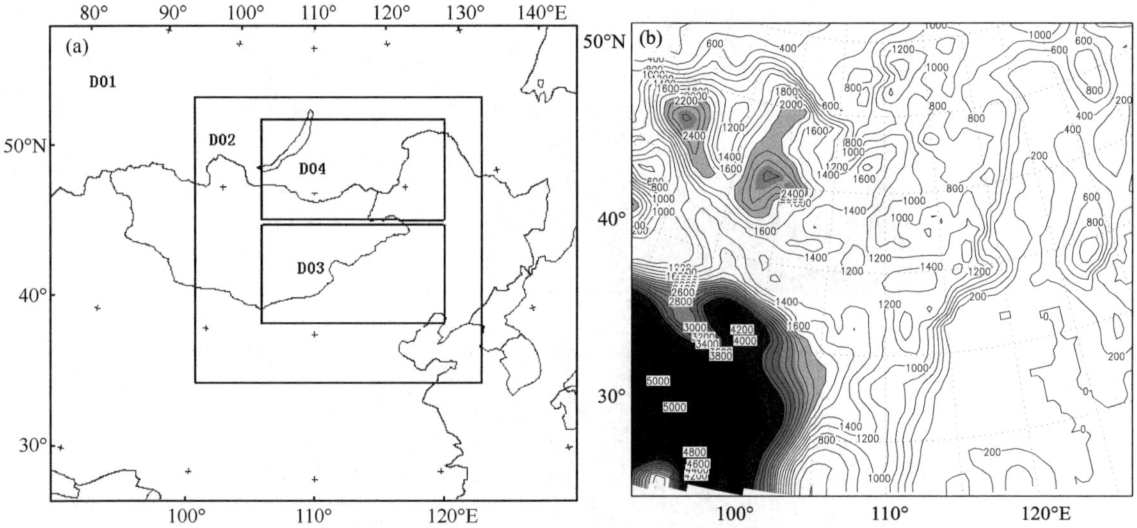

图 2　(a)为模拟区域；(b)Domain1 的地形高度。单位 m，间隔 200。阴影区地形高度大于 2000 m，北部为阿尔泰—萨颜岭山地

4　结果分析

4.1　地形对蒙古气旋各发展阶段界定的影响

图 2b 模拟区域地形高度分布显示，模拟区域地形存在两个明显的高值区：北部范围较小的是阿尔泰—萨彦岭山地，海拔高度锋值约为 4000 m，位于蒙古气旋发生区域的上游，对蒙古气旋发展具有直接的影响；南侧范围很大的是青藏高原，平均海拔高度在 4000 m 以上，它与阿尔泰—萨彦岭山地间形成了明显的峡谷地带，这样的地形分布也将一定程度影响蒙古气旋的

发展。这里,首先分析阿尔泰—萨彦岭山地对蒙古气旋各发展阶段界定的影响。

阿尔泰—萨彦岭山地对东移的低层冷空气产生明显的阻滞(Retard),这一阻滞不同于阻塞(block),阻滞的作用仅仅延缓了低层冷空气的推进。从本次过程 700 hPa 形势演变(图 1)可见,5 日 20 时,由于系统爬坡,等高线明显变平,几乎难以看到高度槽的存在,而温度槽变为西北—东南走向,槽前等温线密集于山地西部(高度与温度这样的配置明显有别于斜压波各阶段的形态),表明低层冷空气受到山地的阻滞。6 日 08 时,这样的阻滞依然存在。相关研究同样表明,这样的阻滞是背风坡气旋发展过程中一个普遍且重要的现象。

另一方面,5 日 02 时,有气旋产生于阿尔泰—萨彦岭山地,此时,低层冷空气未形成阻滞,故气旋呈现典型锋面气旋形态,而其位置决定了此时还不能称为蒙古气旋。5 日 08 时—6 日 02 时(图 1,5 日 20 时),系统爬坡,由于低层冷空气产生阻滞,冷锋消失,锋面气旋变为山地上的一个或多个中低压。6 日 02 时后,低层冷空气越过山地,重新形成锋面气旋,低层斜压强迫开始影响气旋发展。因此,如果以在山地形成低压作为蒙古气旋的开始,则其可以依据是否产生气旋冷锋分为两个截然不同的发展阶段,即:1)触发阶段:冷空气形成阻滞,地面表现为低压。2)发展阶段:冷空气越过山地,地面表现为锋面气旋,低层斜压强迫开始影响气旋发展。可见,蒙古气旋的发展过程可以进行类似 Zupanski 等[4]和 Mattcks 等[5]的阶段划分。

需要指出,在蒙古气旋的触发阶段,由于山地对低层冷空气的阻滞排除了低层斜压强迫对地面气旋发展的直接影响,此时导致地面系统发展主要强迫应该是:1)位涡度守恒导致气柱下坡正涡度的产生。2)Mattcks 提到的“不同阻塞”作用,即低层冷空气阻滞增强了高空急流次级环流产生的地面减压强迫。之后在气旋的发展阶段,斜压强迫成为主要的强迫因素。从本次过程可以看到,气旋在触发阶段的强度、发展速度均较弱。可见,地形影响使蒙古气旋发展过程分为两个截然不同的阶段,在这两个阶段气旋的强度、发展速度、强迫机制存在明显差异。本文主要针对蒙古气旋的发展阶段(6 日 02 时之后)的地形影响进行模拟研究。

4.2 斜压强迫

在蒙古气旋的发展阶段(6 日 02 时—7 日 02 时),按其发展速度明显可以分为缓慢发展阶段和爆发阶段,这还是由于地形对冷空气的阻滞。首先,由于在阿尔泰—萨彦岭南侧形成了峡谷地形,而低空急流位于峡谷中,因此,它能够首先输送冷空气到达背风坡,导致 6 日 02 时地面产生气旋冷锋(6 日 08 时 700 hPa 温度槽呈现西北—东南走向,说明低层冷空气经过峡谷到达背风坡,对流层低层开始产生斜压强迫)。但是,此时冷空气主体依然受到山地阻滞产生堆积,因此,到达背风坡的冷空气势力较弱,斜压强迫的气旋发展速度也相对较弱。6 日 20 时后(图 1),温度槽主体越过山地,表明逐渐堆积的冷空气突然爆发并完全到达背风坡,此时强的斜压强迫导致这一阶段的气旋发展达到了最剧烈的程度(6 日 20—7 日 02 时,平均 1 hPa/h)。可见,斜压强迫是本次过程蒙古气旋发展阶段的主要强迫机制,而地形影响使冷空气经历从缓慢堆积到突然爆发的过程,从而导致蒙古气旋出现从缓慢发展到剧烈爆发的转变。

以下将通过数值模拟进一步分析地形影响斜压强迫进而影响蒙古气旋发展的过程。首先,图 3 为实况和模拟 7 日 08 时(气旋中心气压达到最低 980 hPa)海平面气压场的对比,可见 CE 对本次过程蒙古气旋发展强度、位置的模拟均较为准确。间隔 1 小时的时间演变也显示 CE 模拟出了气旋的整个发展过程。以下将结合模拟结果进行分析。

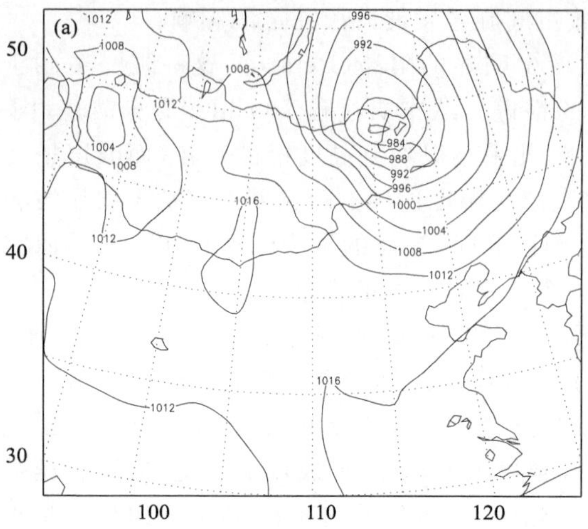

图 3　实况分析的(a)和对照试验模拟的(b)海平面气压(2001 年 4 月 7 日 08 时)。单位：hPa,间隔:4

为了更直接地反映地形对斜压强迫的影响，以下利用倾斜涡度发展理论(SVD)[8~10]进行分析。从 304 K(图 4)等熵面气压分布可以看到，以 110°E 为界，其西侧为低值区，东侧为高值区，说明等熵面在此向下倾斜，最低达 560 hPa，风场显示该区域空气在风场作用下将沿等熵面下滑(向东)。根据 SVD 理论，沿倾斜等熵面下滑的气块将导致静力稳定度的降低，水平斜压度的增强，配合适宜的环境(正涡度区)，垂直涡度将获得迅速增长。

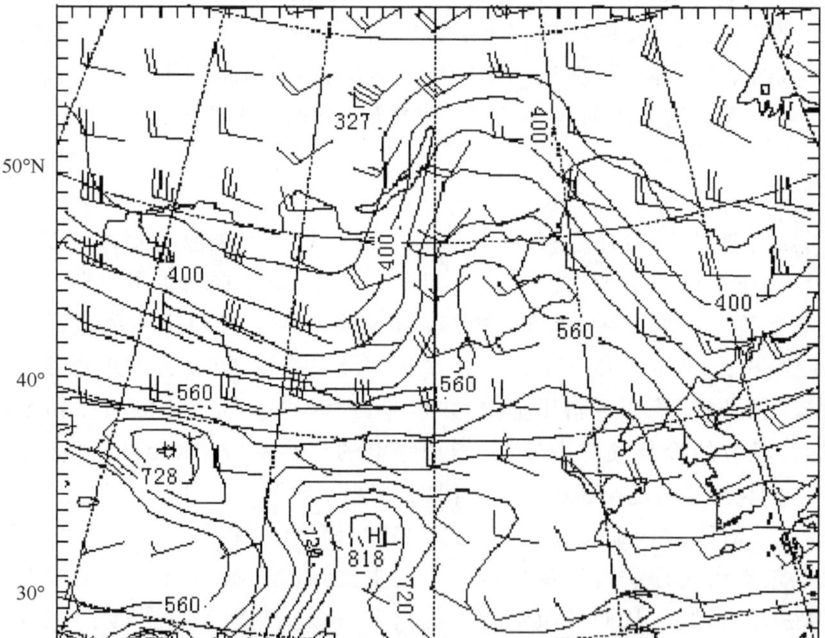

图 4　模拟 9 小时 304 K 等熵面气压场。实线为气压，单位：hPa,间隔:40

在吴国雄提出的全型涡度方程 $F_{cve}=\rho d/d_t(P/\theta_Z-C_D)$ 中(P 为位涡，$C_D=\xi_s\theta_s/\theta_n$)，$C_D$ 集合了多种影响涡度发展的动力、热力因素，用来考察它们对涡度的总体影响。当气块沿倾斜等熵面下滑时，如果 $C_D<0$，且 C_D 逐渐减小时(且该区域为正涡度区)，垂直涡度将得到迅速增长。其中 ξ_s 为垂直风切变产生的水平涡度，θ_s、θ_n 分别为水平斜压度和静力稳定度(可以将沿倾斜等熵面外法线方向的单位矢量 θ 分解为水平、垂直分量，即得到 θ_s 和 θ_n)。

本次过程中 ξ_s、θ_s、θ_n 的分布可通过图 5 看到。图中可见，308 K—288 K 等熵面在阿尔泰—萨彦岭山地东部向下倾斜，且低层更为陡立，经向风 V 也在这一区域分界，将倾斜等熵面外法线方向单位矢量 θ 分解，可知此时 $\theta_s>0$，$\theta_n>0$，而此时垂直风切变项 ξ_s 与 θ_s 反向，即 $\xi_s<0$，因此 $C_D<0$；气块下滑时，θ_s 增大，θ_n 减小，因此 C_D 将减小，垂直涡度将快速增长。可见，斜压作用对气旋发展的强迫通过水平斜压度、静力稳定度、风垂直切变等因素改变得以实现。

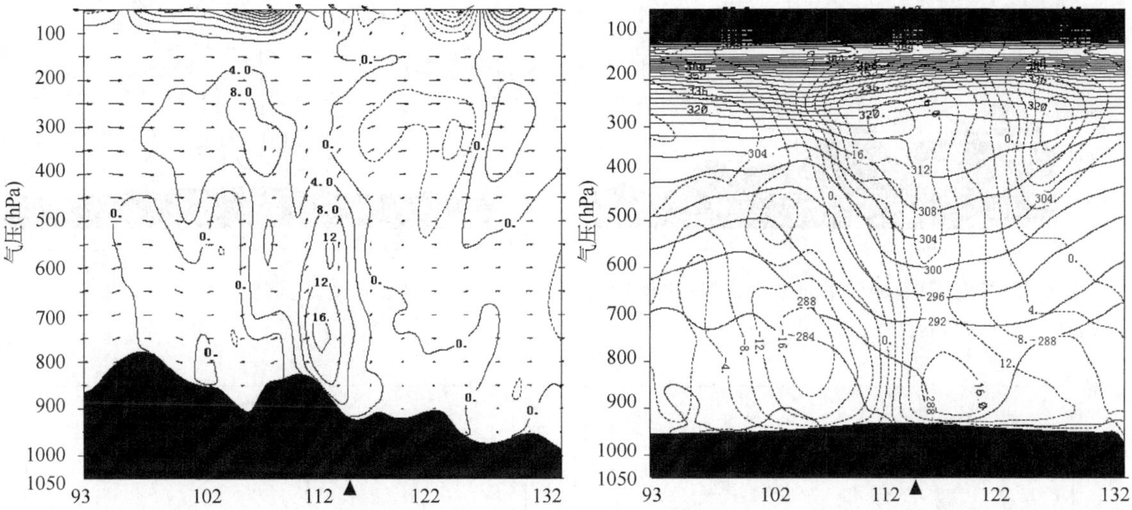

图 5　控制试验(a)和平面地形试验(b)模拟 9 小时穿过气旋中心东西向垂直剖面图。实线为位温，单位：K，间隔：4；虚线为南北风(南风为正)，单位：m/s，间隔：4。三角表示气旋中心

图 6 给出 3 小时后模拟气旋中心区涡度垂直分布，图中可见，此阶段涡度在 400 hPa 以下形成了垂直涡柱，而中心位于 700～800 hPa(16×10^{-5} s^{-1})，与之对应图 5 中等熵面最为陡立的层次也在 700～800 hPa 间，说明在等熵面最为陡立的区域，斜压强迫导致了最强的涡度发展，其强迫的强度与等熵面的陡立程度是正相关的。这里，通过图 5 还可看到地形对等熵面坡度的影响，图中，由于地形对低层冷空气的阻滞，使冷空气在山地迎风坡一侧堆积，使等位温线不能维持其在原先固有的倾斜率，在山地上方变为垂直(使等等位温线不能够随系统东移而在山地上产生截阻)，从而使山地上空等熵面变得更为陡立。根据 SVD 理论，山地的这一影响将明显加剧其上方气柱的斜压作用，从而更利于气旋发展。

为了验证地形对斜压强迫的影响，将 33°N 以北的地形高度取为 500 m，变为平面地形进行试验(PTE)，结果表明：模拟 24 小时气旋强度相差 4 hPa，出现了较明显的减弱(6 日 08 时—7 日 08 时气压变化客观分析为 992～980 hPa，试验 CE 为 992～982 hPa，试验 PTE 为 992～986 hPa)。气旋位置偏北约 3～4 个纬距，表明阿尔泰—萨彦岭山地对气旋形成位置也产生了明显的影响。

对比图 7 与图 5 可以看出：去掉山地地形后，使山地上方的等熵面坡度明显变小，依据有关 SVD 理论的上述分析，等熵面倾斜度越小，斜压强迫的涡度增长也越小。气旋区涡度垂直分布(图 8)显示这一改变导致的结果：平面地形使 CE 中 700～800 hPa 间的涡度中心消失，并

使涡度在对流层中下层呈现均匀分布，说明平面地形使斜压强迫的涡度增长在对流层中低层均匀分布，而山地地形则使斜压强迫的涡度增长向下聚集在紧邻山地的上方。

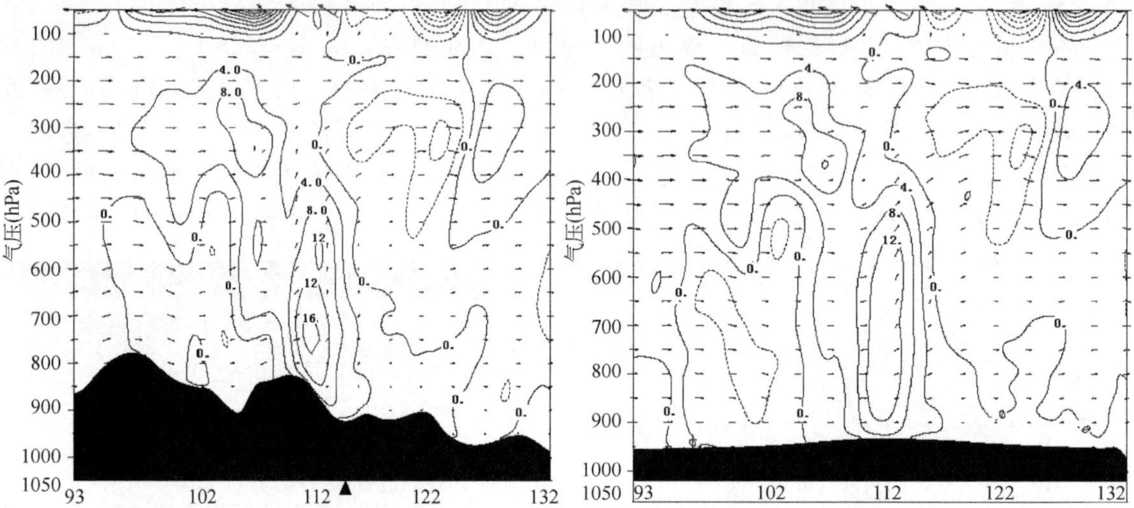

图 6 控制试验(a)和平面地形试验(b)模拟 12 小时穿过气旋中心涡度垂直剖面图。实线为涡度，单位：10^{-5} s^{-1}，间隔：4

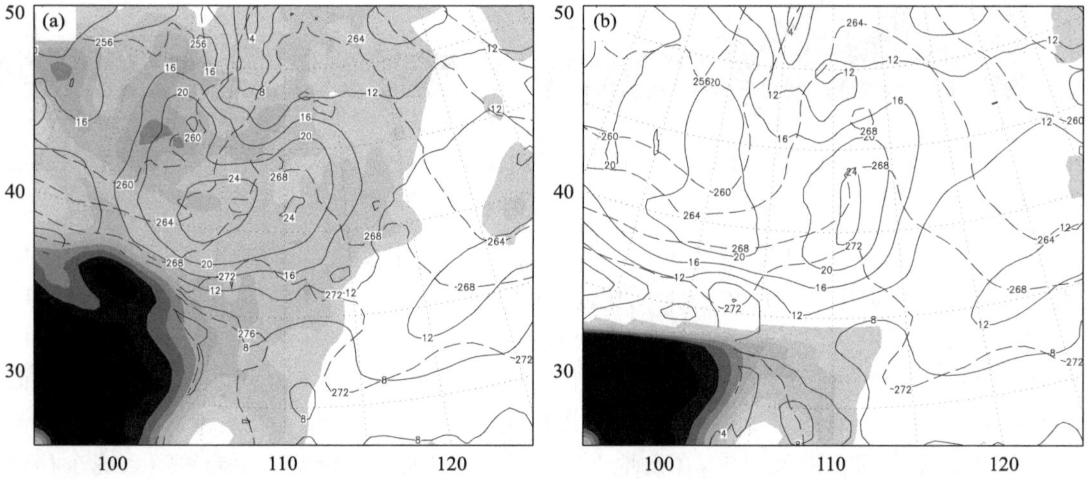

图 7 (a)CE 和(b)PTE 模拟 6 小时 700 hPa 全风速(实线，单位：m/s，间隔：4)和温度(单位：K，间隔：4)。阴影区表示地形高度

斜压强迫高空系统的发展可以视为使该层的涡度获得增长，而高空系统的发展响应到地面引起地面气旋发展的过程，则需要通过涡度平流及差动涡度平流引起的上升运动、辐合辐散等调整过程实现。因此，斜压强迫地面气旋发展是一个间接影响的过程。在本次过程中，山地地形则使(斜压强迫的)涡度增长向下聚集在紧邻山地的上方，即：使低层增强、高层减弱。这一结果显然会导致响应(调整)之后气旋的发展强度强于涡度随高度均匀分布的情况。(这一影响可以形象的看作山地对(斜压强迫)的涡度增长产生向下的“吸引”)。这也是“地形诱发 Terrain induced”的蒙古气旋往往强于“地形影响 Terrain effected”的部分原因(如 2002 年 11 月 10 日形成的一例爆发性蒙古气旋，其高层仅为非常弱的浅槽，而经常出现的在高层形成切

断低压的情况下，地面蒙古气旋发展往往并不能达到爆发性气旋的强度）。

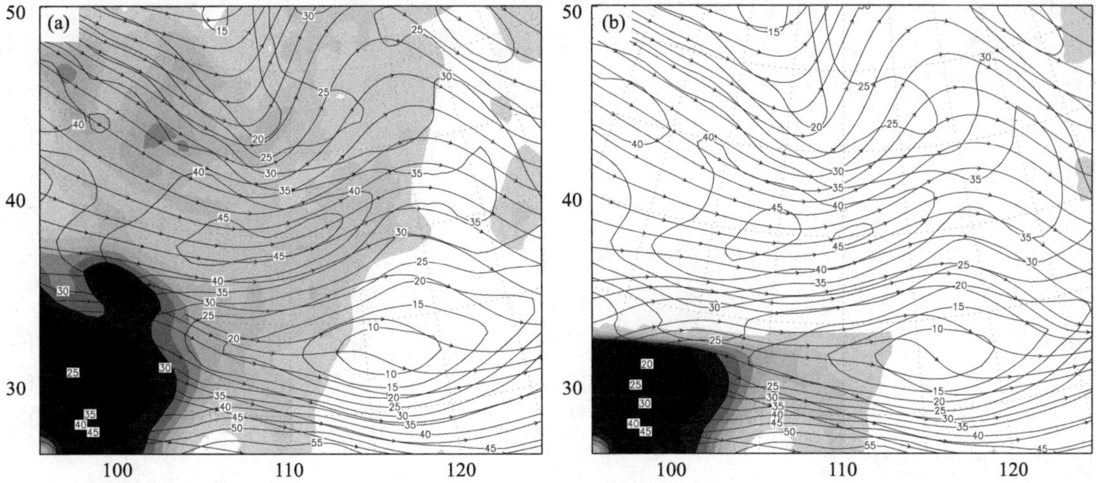

图 8　CE(a)和 PTE(b)模拟 6 小时 250 hPa 全风速(实线，单位:m/s，间隔:5)和流线。阴影区表示地形高度

综上所述，在蒙古气旋发展阶段，地形对斜压强迫产生两种影响。其一，山地地形使冷空气经历一个从缓慢堆积到突然爆发的过程，也使气旋发展经历了一个从缓慢发展到剧烈加强的过程。这一过程相当于将斜压不稳定能量积累起来一起释放，而平面地形情况下斜压能量在这一过程中是相对均匀地释放的，在天气尺度过程摩擦耗散相对较小的情况下，这两个过程将不会导致最终气旋强度的明显差异。其二，山地地形使斜压强迫的涡度增长向下聚集，其显然将使地面气旋发展强度得到明显增强。因此，这一因素应该是山地地形使蒙古气旋强度增强的原因。

4.3　地形的其他影响

本次过程低空急流的演变也明显受到地形的影响。图 1 显示，在阿尔泰—萨彦岭山地与高原间形成了一定的峡谷地形，其产生的直接结果是使低层气流在峡谷入口处汇合，在出口处又疏散，从而在狭谷地区对急流产生强迫。5 日 20 时低空急流位于阿尔泰—萨彦岭山地西侧(急流核在 46°N，42 m/s)，急流南北跨度达 10 个纬距。6 日 08 时低空急流东移到达峡谷区域，其南北跨度出现明显减小（不足 5 个纬距），同时出现一定的南移进入峡谷(急流核在 43°N，24 m/s)。6 日 20 时，急流移过峡谷，南北跨度又开始加大（10 个纬距），强度则明显增强(急流核在 42°N，32 m/s)。可见，峡谷地形对低空急流的南北跨度及位置变化能够产生一定影响。这里，急流强度的变化主要受到山地地形的影响。当系统越山过程中，高度场减弱变平，同时，低层冷空气受到山地阻滞，不能伴随急流东移，都将导致此阶段低空急流强度的减弱。而冷空气越山后，系统增强和强冷平流又将导致其加强。可见，阿尔泰—萨彦岭山地对低空急流的强度变化具有重要影响（峡谷地形也能够对低空急流强度产生一些影响，但影响程度较弱）。

地形也能够对高空急流的演变产生一定的影响，这一影响是通过越山过程中地形导致系统强度的变化引起的。5 日 20 时，系统主体未移上山地（尚未出现减弱），高空急流呈东北—西南向，此时低层锋区（由于地形阻滞位于山地西侧）位于高空急流下方，所以急流较强（急流核为 56 m/s)；6 日 08 时，系统主体移上山地，位于峡谷的低空急流产生冷空气输送使锋区呈现东西向并位于峡谷区域，而高度槽则明显减弱变平，因此，导致高空急流变为东西向且强度出现减弱(急流核 48 m/s)。6 日 20 时，系统主体移过山地，强度迅速增强，高度槽加深，同时，低层冷空气越山爆发使锋区移至背风坡并加强，导致高空急流进一步加强(急流核增至 62 m/s)。这里，高空急流与其下方高空锋区的位置是基本一致的，这反映了热成风平衡对急流位置的约束作用。

以下平面地形试验验证了地形对急流的这些影响。通过图 8 可见，地形改变对低空急流产生了重要的影响，CE 中 6 日 14 时急流为东北—西南向，24 m/s 范围较大；而 PTE 中急流变为南北向，分别位于槽前(向南)和槽后(向北)，且 24 m/s 范围大为缩小。因此，地形对低空急流的影响体现在位置、强度、范围、走向等方面。另外，图 8 显示了地形对高空急流产生的影响。图中平面地形使高空急流范围缩小，急流核出现断裂。但其变化明显小于低空急流的改变，这主要因为地形对高空急流的影响是间接的。高低空急流的这些改变都能够一定程度地间接影响蒙古气旋发展。

必须指出，本文主要针对蒙古气旋发展阶段的地形影响进行模拟研究(6 日 08 时开始)，在此之前，低层冷空气的阻滞、高空槽减弱变平等地形影响都已经出现(5 日 20 时之前)，至 6 日 08 时模拟开始时，地形影响更使高空急流出现明显减弱。这是敏感试验中高空急流改变不很明显的主要原因(另外，模式地形较真实地形有较大的削弱即：4000 m—小于 3000 m，是另一原因)。就蒙古气旋的触发阶段而言，地形影响或许更为重要(作者将另文研究)。尽管如此，可以看到，即使在蒙古气旋的发展阶段，地形依然对蒙古气旋发展强度、低空急流等具有非常重要的影响。这与有关背风坡气旋的研究结论是一致的(地形影响将在整个气旋发展过程中持续)。

4.4 位涡平流对气旋发展的影响

对模拟结果的分析表明，本次过程还存在导致气旋发展的其他强迫因素。从图 6 可以看到，气旋强烈发展阶段涡度在垂直方向存在两个中心，位于 700～800 hPa 的强中心主要源于斜压强迫，另外，在 400 hPa 以上也形成了一个涡度中心(300 hPa，$8\times10^{-5}\ s^{-1}$)位于低涡中心西侧，强度弱于低层。在气旋发展过程中涡度垂直分布始终具有这一特征(在地形影响试验中其未发生改变)，表明其可能源于其它强迫因素。通过对高层位涡的分析，发现其形成源于等熵位涡平流。从 312 K 等熵面(对流层高层)位涡分布(图略)可以看到，气旋发展过程中有位涡大值区向南伸展到地面气旋中心西侧，与高层涡度中心的位置非常一致。从位涡垂直剖面图(图略)可见，高位涡来源于对流层顶和平流层，位涡中心在 150 hPa 左右，随偏北气流向南向下输送，平流主要集中在 200～450 hPa(如果将 2 PVU 作为平流层位涡的标准，则对流顶折叠引起的位涡输送达到 400 hPa)。也与高层形成的涡度中心一致。因此，在高层形成于气旋西侧的涡度中心应该是高层位涡平流产生的，这也是本次过程蒙古气旋发展的一个高层强迫因素。(有关研究表明，在气旋的发展过程中，位涡平流是一种重要的高层强迫因素[11])。

高位涡通常贮存于极地平流层大气中(IPV Reservoir)[11~14]，随对流顶折叠产生向下的输送。对流顶折叠是与高空锋区、高空急流的加强和下落相伴的，可以由高空槽加深、系统加强引发。沈建国等[12]研究表明，平流层等熵位涡能够被平流至气旋中心上空并达到 650 hPa。陈受钧等[1][2]的研究显示，源于高层强迫的蒙古气旋首先在对流层高层出现切断低压，并形成高空槽前的西南急流使位涡平流能够到达地面气旋中心上空。而本次过程显然源于低层斜压强迫，且未形成西南向的高空急流，故位涡平流未到达地面气旋中心上方，且影响程度也相对较弱。

5 总结

本文通过对影响蒙古气旋发展的地形因素进行的天气分析及针对其发展阶段进行的数值模拟研究，结果表明：

(1)地形对蒙古气旋的影响导致蒙古气旋可以分为两个截然不同的发展阶段，即：触发阶段和发展阶段。它们本质的差异是低层冷空气是否直接导致地面气旋冷锋产生，即其中斜压强迫的重要程度。

(2)本次过程蒙古气旋的发展阶段，其强迫主要来自斜压不稳定。地形的影响在于：1)阿尔泰—萨彦岭山地对低层冷空气产生明显的阻滞。这种阻滞使冷空气经历从缓慢堆积到突然爆发的过程，从而导致蒙古气旋出现从缓慢发展到剧烈爆发的转变。2)阿尔泰—萨彦岭山地对低层冷空气的阻滞使山地上空等熵面更为陡立，(依据 SVD 理论)从而明显加强对流层低层的斜压强迫。这种加强的本质是使斜压强迫的涡度增长向低层聚集，这一影响将导致地面气旋发展的发展强度获得明显的增强。

(3)阿尔泰—萨彦岭山地及其与青藏高原间形成的峡谷地形对低空急流的位置、范围、强度及演变过程产生影响。另外，地形通过对系统强度演变的影响间接影响高空急流。并通过它们的改变影响气旋发展

(4)对流层高层位涡平流的输送导致在 400 hPa 形成偏于气旋西侧的涡度中心。位涡的这一输送未能直接到达气旋中心上方，且强度较弱。

参考文献

[1] Chen S J, Egger J and Han W. Altai-Sayan lee cyclogenesis: A case study. *Beitr. Phys. Atmos.*, 1992, **65**: 13-22.

[2] Han W, Chen S J and Egger J. Altai-Sayan lee cyclogenesis: Numerical simulations. *Meteorl. Atmos. Phys.*, 1995, **55**: 125-134.

[3] Buzzi, Tibaldi A S. Cyclogenesis in the lee of the Alp: A case study. *Quart. J. Roy. Meter. Soc.*, 1978, **104**: 271-287.

[4] Zupanski and Mcginley J A. Numerical analysis of the influence of jets, fronts, and mountain on Alpine lee cyclogenesis. *Mon. Wea. Rev.*, 1989, **117**: 154-176.

[5] Mattocks C, and Bleck R. Jet streak dynamics and geostrophic adjustment processes during he initial stages of lee cyclogenesis. *Mon. Wea. Rev.*, 1986, **114**: 2033-2056.

[6] Bleck, R. and Mattock. A preliminary analysis of the role of potential vorticity in Alpine lee cyclogenesis. *Contrib. Phys.* 1984, **11**: 38-69.

[7] Grell G A, Dudhia J and Stauffer D R. *A description of the fifth-generation Penn State/NCAR mesoscale model(MM5)*, 1995, NCAR/TN-398+STR.

[8] 吴国雄，蔡雅萍，唐晓晴. 湿位涡和倾斜涡度发展. 气象学报，1995，**53**: 387-405.

[9] 吴国雄,刘还珠.全型涡度倾向方程和倾斜涡度发展.气象学报,1999,**57**:1-15.

[10] 吴国雄.全型涡度方程和经典涡度方程的比较.气象学报,2001,**59**:384-392.

[11] Thorpe A J. Diagnosis of balanced vortex structure using potential vorticity. *J. Atmos. Sci*, 1985, **42**: 397-406.

[12] 沈建国等.1993,爆发性蒙古气旋若干统计特征.内蒙古气象,1993,**4**:5-12.

[13] Karyampudi M V, Kaplan M L, Koch S E and Zamora R J. The influence of the Rocky Mountains on the 13-14 April 1986 severe weather outbreak. PartI: Mesoscale lee cyclongenesis and its relationship to severe weather and dust storms. *Mon. Wea. Rev.*, 1995, **123**: 1394-1422.

[14] Mcginley J A, Goerss J S. Effect of terrain height and blocking initialization on Numerical simulation of Alpine lee cyclongeness. *Mon. Wea. Rev.*, 1986, **114**: 1578-1590.

Observational and Numerical Study of Topography Influence on Mongolia Cyclogenesis

Jiang Xuegong[1*] Shen Jianguo[2] Liu Jingtao[1] Chen Shoujun Wang Yingchun[4]

1. Inner Mongolia Meteorological Observatory, Huhhot, China;

2. Inner Mongolia Meteorological Bureau, Huhhot, China;

3. Department of Atmospheric Sciebces, Peking University, Beijing China;

4. Beijing Urban Meteorological institute, CMA, Beijing

Abstract Observational study and numerical simulation are conducted on the topography influences on the Mongolia cyclogenesis. The results shown that the Mongolia cyclogenesis can be divided into two processes, namely spring process and development process, according to the appearance of baroclinic forcing in lower troposphere. In the development process, the baroclinicity was the main forcing mechanism causing the development of Mongolia cyclone. The topography influences focus on: 1) The Altai-Sayan complex mountains retarded the cold air in lower troposphere and so the Mongolia cyclone developed slowly before the cold air completely struck the lee side and after that it developed explosively. 2) The baroclinicity was enhanced by the complex mountains (steepened the isentropic surface), which made the enhancement of vorticity concentrated downward at the low troposphere and so reinforced the development of surface cyclone. 3) The complex mountains affected the position, range and strength of lower level jet.

In the upper troposphere the isentropic potential vorticity (IPV) advection contributed to the cyclogenesis and its effect was limited.

Key words: Mongolia cyclogenesis, Topography influences, Numerical simulation

地形影响沙尘传输的观测和模拟研究*

姜学恭[1] 陈受钧[2]

1. 内蒙古自治区气象台，呼和浩特 010051；2. 北京大学物理学院大气科学系，北京 100871

摘 要：本文基于观测资料和数值模拟方法，研究了东亚地形对我国沙尘传输的影响，结果表明：东亚地区沙尘天气多发区主要位于我国南疆盆地和内蒙古西部及蒙古南部。南疆盆地沙尘天气集中在盆地南缘；而蒙古、我国内蒙古西部主要出现在沙漠腹地。青藏高原东北侧到黄土高原中部是沙尘天气次多发区。蒙古、我国内蒙古西部沙漠腹地多发区的形成不仅由于这里提供了丰富的沙源，同时也具备了沙尘暴迅速增强的条件：阿尔泰—萨彦岭南侧的峡谷地形强迫形成峡谷急流，明显增强了该区域地面风速；萨彦龄山地南坡携带大量沙尘的向南下坡气流遇到东—西向的峡谷气流时受到阻挡形成聚集；同时，这一地区起伏的地表产生的地形波加强了地面起沙。这种大量沙尘在大气中聚集之后再进行传输的特征可以视为在大气中形成了沙尘"中继站"。导致沙尘进一步向东输送的地形因素是阿尔泰—萨彦岭山地南侧的峡谷地形，而萨彦岭山地南坡的下坡气流和青藏高原东北侧的地形强迫绕流是导致沙尘向南输送的原因。由于青藏高原地形绕流的强大以及冷锋过程的相对频繁使得沙尘的向南输送更为强盛，这也是青藏高原东北侧沙尘天气多发区的形成原因。

关键词：沙尘暴，地形，数值模拟

1 引言

近年来，亚洲频繁发生的沙尘暴事件引起了世人的广泛关注，沙尘暴不仅频频出现在沙尘源地及其周边地区，同时也通过远距离传输过程到达遥远的下游地区，给这些地区的工农业生产、交通运输、人民生活和身体健康及生态环境造成巨大的危害。据有关研究估计[1]，每年约有 1.0～3.0 Gt 的沙尘粒子注入大气，而在亚洲仅春季向大气中排入大气的沙尘就达120 Mt。据 Zhang，X Y、Gong，SL、Zhao，TL[2~4]等基于数值模拟并结合观测资料对亚洲沙尘排放的研究表明，在春季从东亚到北美西部的中纬度对流层中绝大部分沙尘气溶胶来自亚洲，而蒙古及我国北方的沙漠区（在我国主要为塔克拉玛干沙漠、巴丹吉林沙漠等）则是亚洲沙尘气溶胶的主要来源，亚洲沙尘气溶胶约 70％来自这些地区。这些沙尘大约半数（51％）在沙尘源区形成沉降，其余部分则被气旋冷锋等天气系统扬升至自由大气（3～10 km）并主要沿 40°N 附近的带状区域向下游输送，其中 21％在亚洲内陆沉降，9％在太平洋沿岸沉降，约 16％在太平洋中沉降，另有 3％跨越太平洋到达北美大陆沉降。Hacker 等[5]就曾对 1998 年 5 月形成于蒙古国西部的"越洋"沙尘暴过程进行了分析和模拟研究。

从某种角度来说，沙尘暴之所以广受关注是因为沙尘的远距离输送。且不论沙尘源区和下游地区在人口密度、生产力水平、经济发展程度等方面的差异，仅就沙尘暴的影响范围而言，沙尘暴在源区形成后通过远距离传输使其影响区域呈百倍的增长，其对人们生产、生活的影响

* 资助课题：国家自然科学基金项目（40565003），科技部社会公益研究专项（2005DIB3J108）资助。

及对气候、环境及生态的危害也成百倍的增长。因此,沙尘暴事件更应该称为"风送沙尘"过程。就沙尘暴的治理而言,更应该深入了解沙尘传输规律,这样才能放眼全局而不是囿于局部,防治沙尘暴才能事半功倍。

影响沙尘传输的根本因素是天气系统[6],它决定了沙尘输送的距离、方向、影响的范围等。更直接的影响因素是沙尘层的主导气流,这在地形较为平坦的地区往往表现的非常明显,如撒哈拉沙漠的沙尘往往在赤道东风的作用下向大西洋传输。但在地形高度较高和地形起伏较大的地区,沙尘的传输很大程度受到地形的影响[7]。亚洲沙尘往往上扬至对流层中上层,是与亚洲沙尘源区多为海拔较高的高原、山地有密切的关系[7]。另外,从沙尘源区到下游总的地形特点是西高东低、西北高东南低(上游高,下游低),因此,发生在沙尘源区的沙尘天气很容易沿向东、向南方向向下游扩展,到达东亚广大平原地区及太平洋沿岸,甚至能够跨越太平洋影响美洲西部。然而,对于东亚地形影响沙尘传输的问题,国内外还较少专门的研究。因此,本文拟通过历史统计、观测分析、个例研究与数值模拟等方法初步探讨东亚地形对沙尘暴传输过程的影响。

2 东亚地形特点及对沙尘天气的可能影响

图 1 为东亚地区沙尘天气年平均日数分布图,图中沙尘天气包括浮尘、扬沙、沙尘暴。使用资料为中国气象局国家气象信息中心整编的全球沙尘天气地面观测资料,资料起止时间为 1980 年 1 月 1 日至 1999 年 12 月 31 日(由于 2000 年后该资料仅包括沙尘暴资料,而有无该资料对本研究影响不大,故本文未采用),每日 02、08、14、20 时四次观测。统计的频数为沙尘天气发生日数,即每天只要有一个时次出现沙尘天气,则计为一个沙尘天气日(统计沙尘天气日数对研究沙尘天气的影响和危害更为适宜,一些研究对沙尘天气过程的次数进行统计,但更有利于进行天气学分析)。

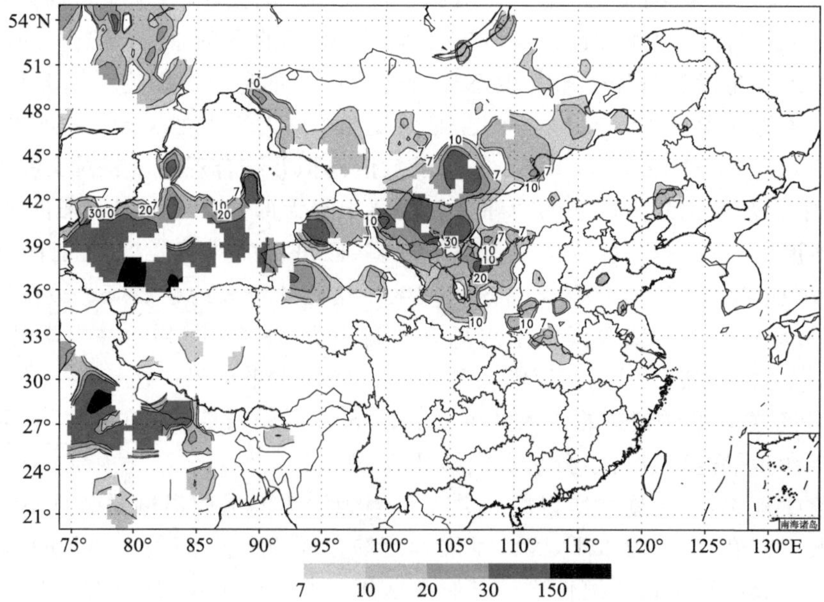

图 1 东亚地区沙尘天气年平均日数(单位:d)分布图

图 2 为东亚地区地形高度分布图。对比图 1 和图 2 可以看到，在东亚有两个明显的沙尘天气多发区，一个位于蒙古南部、我国内蒙古西部，年平均沙尘天气次数 20～50 次，明显高于周围其他地区。另外一个在南疆盆地，年发生次数在 30 次以上，盆地南部边缘达到 100 次以上。它们也是影响东亚的主要沙尘源区。但是，这两个地区沙尘天气的影响是不同的。发生在南疆盆地的沙尘天气尽管非常频繁，但较少能够移出盆地，或移出盆地但造成的影响范围较小。而发生在蒙古南部、我国内蒙古西部的沙尘天气几乎每次都能够完成向东、向南的输送，正如前文提到的，其影响甚至到达太平洋东岸。分析它们的地形特点，也许能够发现一些端倪。首先，这两个地区有一些共同的地形特点，南疆盆地位于天山山脉和青藏高原（昆仑山、阿尔金山）之间的低洼地带（图 2）；而蒙古南部和我国内蒙古西部沙漠区则位于萨彦岭山地与祁连山之间广阔的低洼地带。不同的是南疆盆地西侧为帕米尔高原，因此三面环山，向东的出口被天山西北—东南向的支脉与西南—东北走向的阿尔金山脉强迫形成向东的倒喇叭口地形。因此，春季由北向南（或沿西北—东南方向）推进的天气系统在越过天山到达盆地之后，相对难以越过盆地四周山地的阻挡（因为沙尘天气主要发生在边界层中），形成沙尘天气的扩展，当然，极小部分的沙尘能够进入狭窄的河西走廊向东传播，但影响范围很小。而蒙古南部、我国内蒙古西部沙漠区的北侧为西南—东北走向的萨彦岭东部支脉，南侧为西北—东南走向的青藏高原东北坡（阿尔金山脉），西部入口则是准噶尔盆地（阿尔泰山与天山之间）的出口，因此，形成与南疆盆地正好相反的正喇叭口地形。蒙古、我国内蒙古西部沙漠区正好位于“喇叭”的中段。因此，在这一区域形成的沙尘天气无疑有利于向东、向南的扩展。更为重要的是，尽管蒙古、我国内蒙古西部处于相对低洼地带，但从萨彦岭山地沿西北—东南方向到蒙古高原直至黄土高原和华北平原，基本维持一个缓慢的下坡过程。而春季夹带大量沙尘的冷空气南下爆发路径往往就是沿西北—东南方向。这样发生在边界层中的沙尘天气很容易沿西北—东南方向扩展。

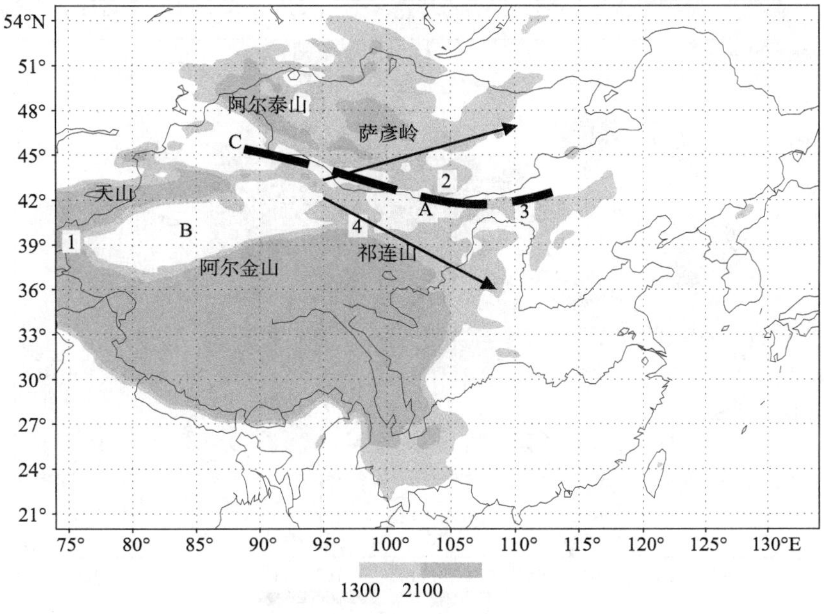

图 2　东亚地形高度分布图（单位：m）

其中，1. 为帕米尔高原，2. 为蒙古境内的戈壁阿尔泰山，3. 为内蒙古中部阴山山脉，4. 为河西走廊入口。A 为蒙古南部、内蒙古西部沙漠区，B 为塔里木盆地，C 为准噶尔盆地可见，从萨彦岭沿东南方向的下坡地形为沙尘天气的扩展提供了有利条件。因此，下文将基于一次影响我国的大范围沙尘天气过程，利用数值模拟和地形敏感性试验，进一步研究上述地形对沙尘天气发生和扩展的影响。

3 天气过程概述

2002 年 3 月 18—22 日沙尘暴过程是近些年影响我国的一次较强沙尘暴过程，从图 3a 本次过程全国沙尘天气分布图可以看到。我国北方大部分地区都出现了沙尘天气，其中，沙尘暴

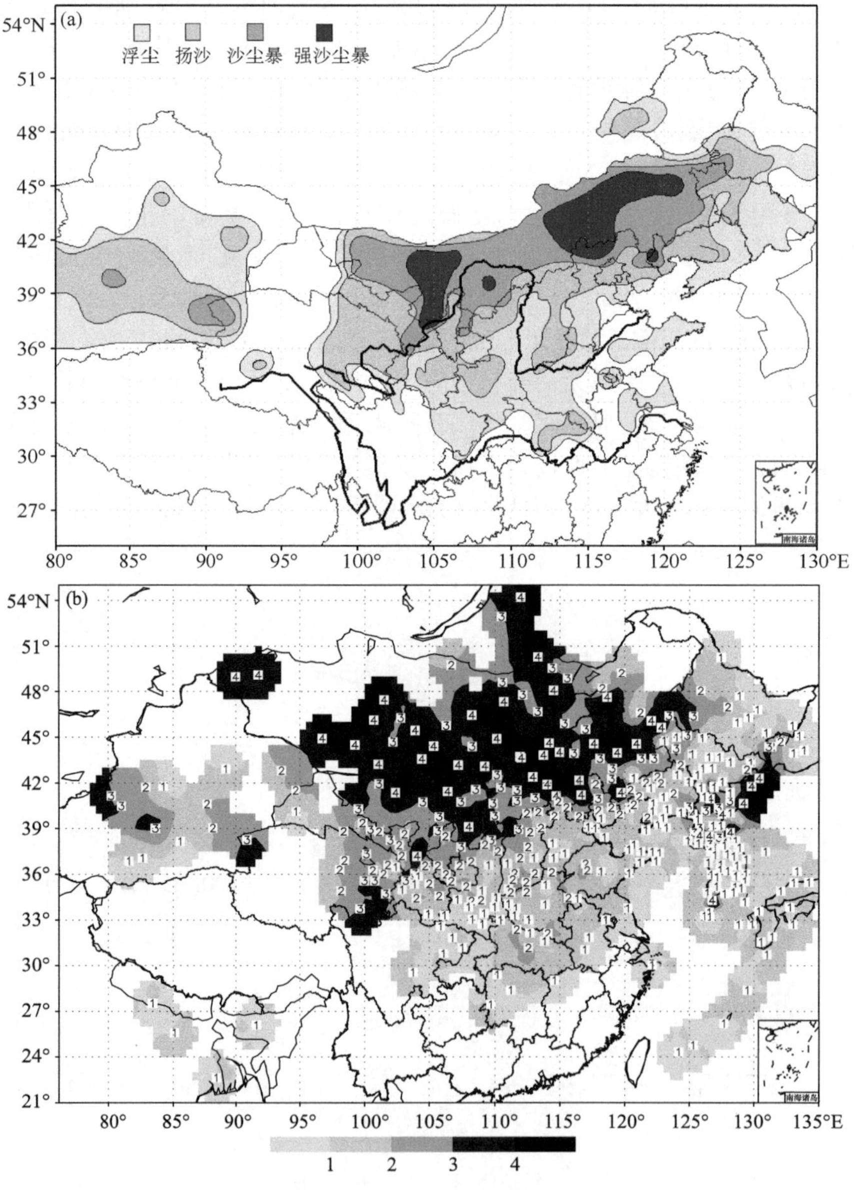

图 3 2002 年 3 月 18—22 日我国(a)和东亚(b)沙尘天气分布图

(1,2,3,4 及不同灰度分别代表浮尘、扬沙、沙尘暴和强沙尘暴)

主要出现在内蒙古地区，并存在两个明显的强沙尘暴中心，一个位于内蒙古西部沙漠区，另外一个位于内蒙古中部、东部偏南地区。这两个强沙尘暴区占到全国沙尘暴区的90%以上。但同时，河北、北京、吉林等地也出现了沙尘暴，而浮尘和扬沙天气更是扩展到了中原广大地区。图3b为东亚沙尘天气分布，资料来自常规地面观测。图中国内部分与图3a利用加密观测统计的我国沙尘天气区相当一致。国外部分在蒙古国出现了非常强的沙尘暴，在西太平洋及其沿岸地区则出现了大范围扬沙天气，从韩国、日本向南伸展到我国台湾附近并呈明显带状分布，显示沙尘到达韩国后进一步向南输送。另外，沙尘天气总的分布呈较明显的涡旋状且主要向东扩展，表现出蒙古气旋的重要影响。

本次过程是一次较典型的蒙古气旋型沙尘暴过程。18日08时，对流层中层(500 hPa)有一低槽在中亚地区生成，之后，低槽东移加深，19日08时到达我国新疆东部(95°E)，受其影响，在阿尔泰—萨彦岭山地东部产生蒙古气旋，扬沙和沙尘暴天气开始出现在蒙古国中西部。19日20时，低槽进一步东移加深，冷平流加大，同时，地面蒙古气旋进一步加强，气旋冷锋开始侵入内蒙古。沙尘暴出现在内蒙古西部及我国西北地区。20日，高空锋区迅速增强并推进至河套一带，同时，地面气旋东移并迅速增强，气旋冷锋扫过我国北方大部地区，带来强烈的沙尘天气。21日，气旋东移，气旋冷锋进入东北地区，沙尘暴区也推进至吉林、辽宁等地。之后，气旋减弱东移，影响我国大部地区的沙尘暴过程也逐渐结束。而浮尘天气继续东移入海，影响更广阔的地区。

4 模式简介和模拟方案设计

研究采用的模式为北京城市气象科学研究所开发的耦合的沙尘预报模式[8]。天气模式为中尺度模式MM5V3[9]。与之耦合的沙尘传输和沉降模式采用与MM5相同的σ坐标，沙尘粒子的水平扩散、垂直扩散及差分方案与MM5相同。为了精确反映沙尘粒子的尺度特征，将沙尘粒径分为16档。模式的起沙过程利用参数化方法，与地表类型、植被覆盖度、土壤湿度等因素有关。是否产生起沙则由摩擦速度决定。起沙后，网格可分辨的沙尘上扬由上升运动和沙尘粒子的沉降速度决定，次网格则由对流调整决定。沙尘沉降分为干沉降和湿沉降，大粒子的干沉降主要为重力沉降，小粒子沉降由湍流扩散和重力沉降共同确定。显式湿沉降由降水粒子对沙尘粒子的冲刷决定，隐式湿沉降由降水量决定。

本文模拟采用T213分析和预报场分别作为初始场和侧边界。模式格距30 km，格点数131×121，模拟起始时间3月19日08时(北京时，下同)，模拟时间48小时。控制试验采用实际地形(图4a)。地形敏感性试验中，对阿尔泰—萨彦岭山地及其南坡地形高度进行了降低(图4b)。调整方法为对上述地区的地形高度进行平滑，使地形峰值从2500 m左右调整到1500 m左右。以下将利用模式输出结果对地形影响沙尘暴的主要阶段(19日08—20日08时)进行进一步的分析。

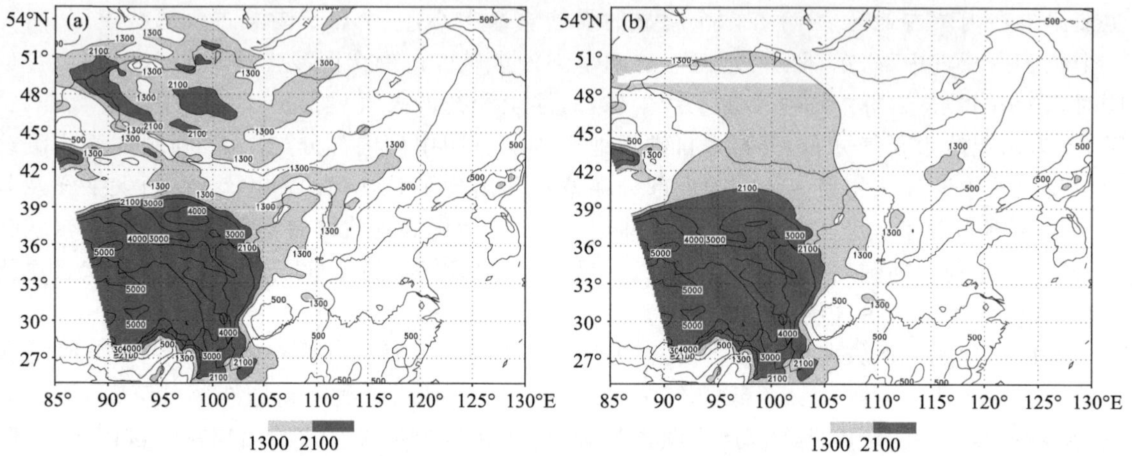

图 4 控制试验(a)和地形敏感试验(b)模式地形高度。单位:m,间隔:400

5 结果分析

5.1 对沙尘天气的模拟

图 5 分别为 3 月 19 日 14 时、20 时及 20 日 02 时、08 时地面观测的沙尘暴强度和控制试验模拟的地面沙尘浓度。可以看到模式对整个沙尘暴发生和扩展过程进行了较好的模拟。19 日 14 时,观测显示大范围的沙尘暴出现在内蒙古与蒙古交界的荒漠化地区,模拟沙尘浓度也在这一地区出现大值区,与实况有着较好的对应。19 日 20 时,观测的沙尘暴增强,站数增加,同时向东有所扩展,而模拟的沙尘浓度也明显增强,从位置上看向东有所推进。20 日 02 时,沙尘区继续向东推进,而模拟的沙尘高浓度区与之有着很好的对应。至 20 日 08 时,观测显示沙尘区的移动出现了明显的分支,向东已经推进至内蒙古中部和东部偏南地区,强度较强,而向南也出现了一个明显的输送带,到达了陕西和山西的大部分地区。而模拟的沙尘高浓度区也分别向东和向南扩展至上述两地,其中,内蒙古中东部偏南地区强度较强,陕西和山西等地稍弱。可见,模式较好地模拟出了本次过程沙尘暴的发生和扩展过程。

图 5 也显示模拟结果存在一些不足,19 日 14 时,模拟沙尘浓度较地面观测沙尘暴强度略偏弱,因为此时积分仅 6 小时,因此模式的 spinup 效应对沙尘强度具有一定的影响。

5.2 沙尘暴发展阶段

从图 5 还可看出,在不同阶段沙尘天气演变过程具有不同的特点:19 日 14 时—19 日 20 观测显示沙尘暴区主要局限在蒙古南部及我国内蒙古西部(尽管沙尘暴区向东略有扩展),同时 14 时在蒙古国西南部发生的沙尘暴到 20 时仍然在持续,而我国内蒙古西部出现沙尘暴的测站明显增多,因此,该阶段沙尘天气主要特点是沙尘暴增强。这一点在模拟结果中有着更明确的反映。之后特别是 20 日 02 时—20 日 08 时,从图 5c,d 可以看到蒙古西部、我国内蒙古西部的沙尘暴强度明显减弱或停止,同时主体沙尘区向东的推进和向南的扩展明显加强(特别是向南的扩展)。因此,沙尘区的扩展成为该阶段沙尘天气演变的主要特点。

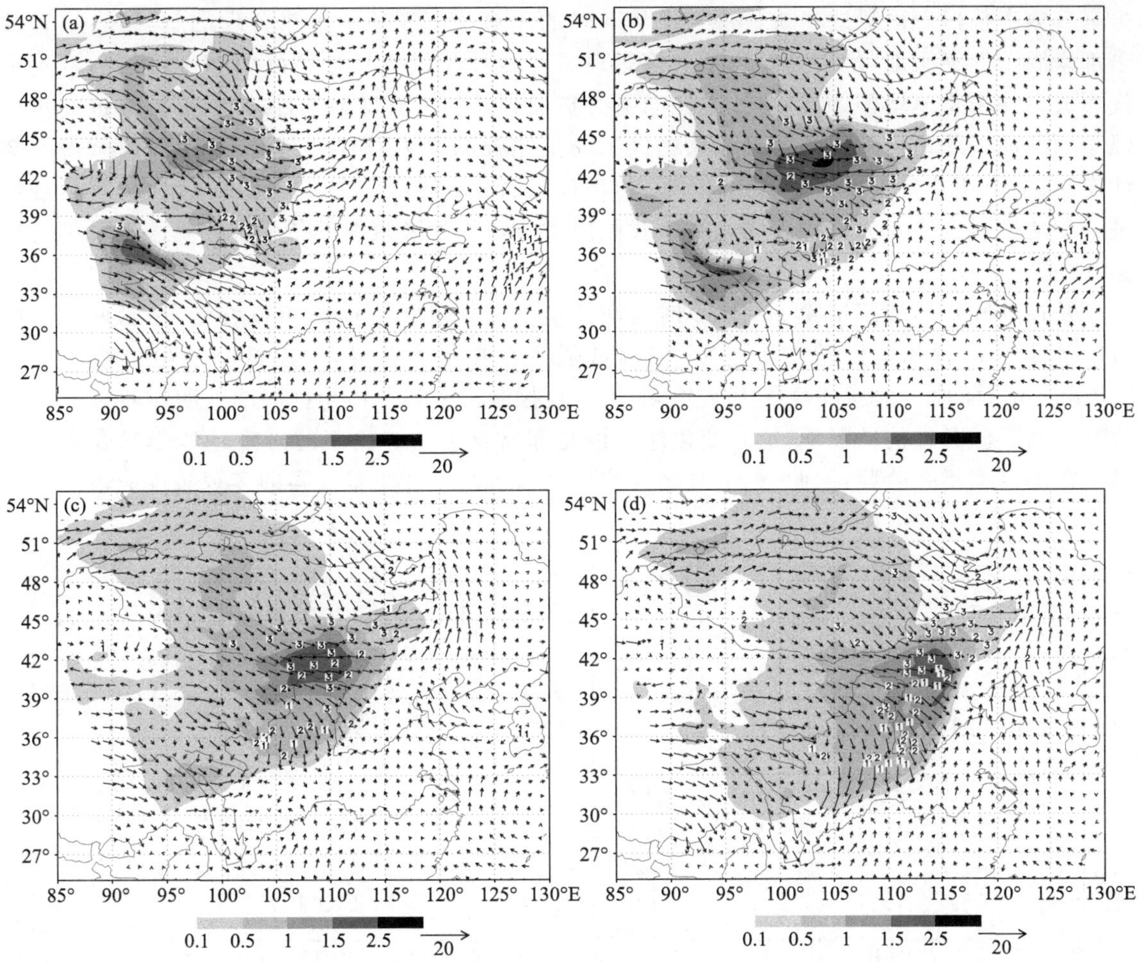

图 5 控制试验 3 月 19 日 14 时(a)19 日 20 时(b)20 日 02 时(c)20 日 08 时(d)模拟地面沙尘浓度(阴影区,mg/m^3)、地面 10 m 风场和地面观测沙尘天气强度(1 为浮尘、2 为扬沙,3 为沙尘暴)。

上述特点与我国北方沙尘暴的一个重要规律是吻合的,即发源于蒙古、我国内蒙古沙源地的沙尘暴总是在该地区迅速增强之后,才开始向东或向南扩展。其下游的沙尘天气则来源于沙尘暴的扩展。这一表述容易使人简单地理解为"这一地区是沙尘源地"。事实上同样作为沙尘源地,不同地区情况是不同的。再回到图 1,可以看到在蒙古及我国内蒙古西部沙漠腹地沙尘天气发生次数明显要多于其南北两侧(尽管蒙古国常常出现缺测造成资料不连续,但这一特征还是非常明显)。而南疆盆地塔克拉玛干沙漠腹地沙尘天气次数却明显少于盆地南缘,与盆地北缘也只大致相当。尽管沙漠腹地测站较少,但接近沙漠腹地的巴楚站(51716,78°34′E,39°48′N)为 33 次,铁干里克(51765,87°42′E,40°38′N)为 28 次,沙漠腹地的塔中站(51747,83°40′E,39°N)从 1999 年才有连续的沙尘天气观测,从 1999 年至 2002 年四年间共出现沙尘天气 96 次,年均 24 次。而沙漠南缘均在 100 次以上。可见,尽管 1999 年至 2002 年为沙尘天气多发期,但塔中年均沙尘天气次数仍远小于盆地南缘。从钱正安[10]等给出的 1952—2001 年我国北方沙尘天气分布图和 1964—1998 年中国北方、蒙古沙尘天气分布图也可看出南疆盆地南缘的沙尘暴出现次数明显高于盆地腹地,而蒙古西部、我国内蒙古西部沙漠腹地反而是沙尘暴日数高值区。可见这一统计事实还是准确的。对这一现象的一个合理的解释是：尽管沙尘都来源于沙漠腹地，但塔克拉玛干沙漠腹地不具备沙尘天气迅速增强的条件，因此，沙尘天气形成后并未迅速增强，而是在沙尘区向下游扩展到具备了其增强条件的地区时，才进一步增强。而盆地南缘地形强迫的抬升显然提供了使其增强的动力条件。在蒙古、我国内蒙古西部沙漠腹地，由于具备了使沙尘暴迅速增强的条件，因而沙尘天气总是在此迅速增强，然后进行进一步地扩展。下文分析将表明，蒙古、我国内蒙古西部沙漠腹地具备了这些使沙尘暴迅速增强的条件。

从控制和地形敏感试验 19 日 20 时地面沙尘浓度分布(图 6a，b)可以看到，尽管两试验模拟的沙尘区位置和范围基本上一致，但两者强度存在巨大差异，控制试验沙尘中心强度为 2.8 mg/m^3，而地形敏感试验仅为 1.2 mg/m^3，且控制试验整个沙尘区的浓度明显较强。在未改变地面起沙条件的情况下，这一变化应该说是非常巨大的。由于此阶段是沙尘暴在进行扩展之前的迅速增强阶段，因此，前文提到的这一地区有利于沙尘暴增强的客观条件应该是地形因素。

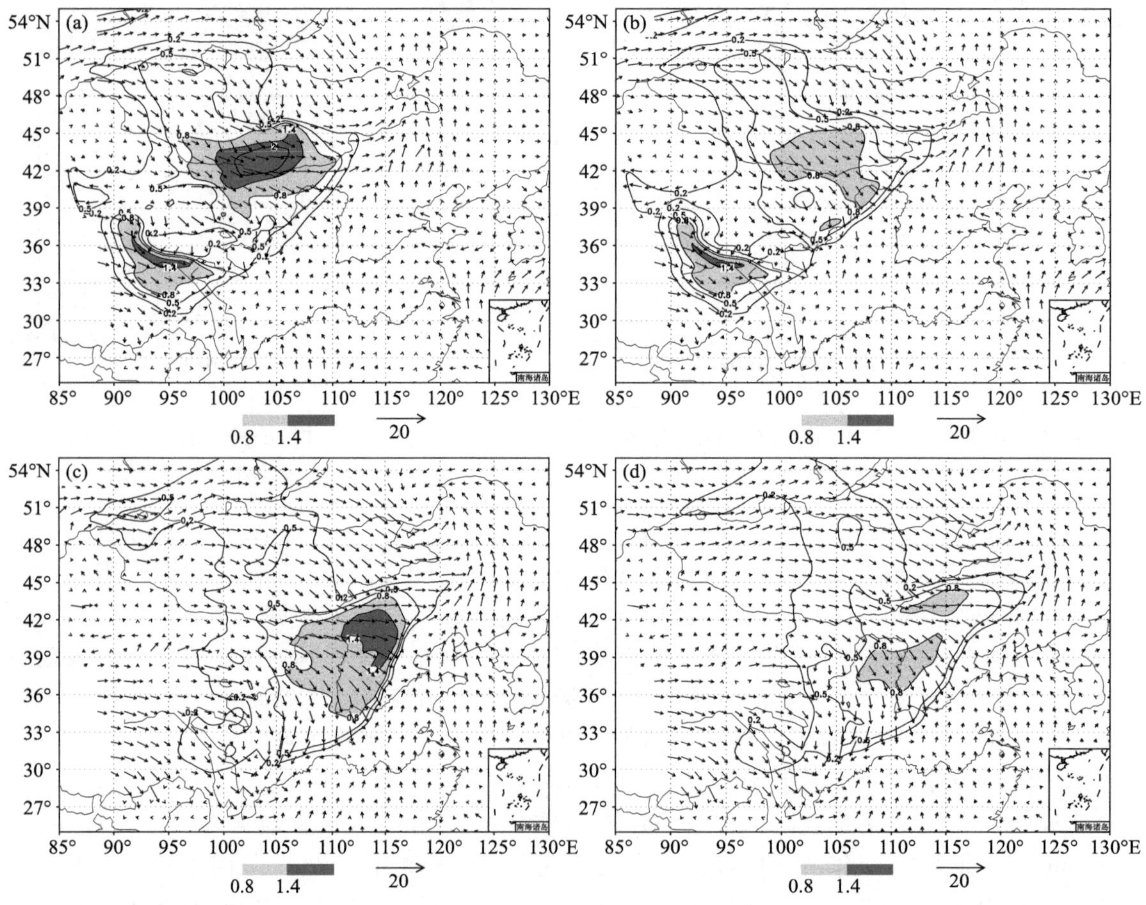

图 6　3 月 19 日 20 时控制试验(a)和地形敏感试验(b)及 3 月 20 日 08 时控制试验(c)和地形敏感试验(d)模拟地面沙尘浓度(mg/m^3)和地面 10 m 风场。

图 7a,b 为两试验地形与地面 10 m 风场分布。进一步分析该地区地形特点可以发现：在向东敞开的喇叭口地形的中部，绵延分布着两条准东西走向的山脉，一个是位于蒙古南部的戈壁阿尔泰山，一个是位于我国内蒙古中部的阴山山脉(图 2)。这两条山脉之间形成了明显的峡谷地形，并与阿尔泰—萨彦岭山地南侧喇叭口附近的峡谷连接，形成了绵延近 2000 km (90°—110°E)的狭长的峡谷地带。这一峡谷地形对低层气流产生了很大的影响。从图 7a 控制试验的地面 10 m 风场可以看到，在峡谷中，气流被峡谷约束，变为与峡谷走向严格一致的准东西向的“峡谷气流”。而其南北两侧则是明显的西北气流。而在敏感试验中，这一区域气流为一致的西北气流。故峡谷地形首先改变了大气低层风场的方向。另外，峡谷地形具有的一个重要作用是导致气流遇到山壁形成绕流，从而使气流在峡谷中增强形成“峡谷急流”，从图 8 控制试验与地形敏感试验地面 10 m 风速差值可以看到，就在峡谷所在区域，控制试验比地形试验风速强 2～4 m/s 左右，而其周围风速改变很小。可见，该峡谷地形也使蒙古、我国内蒙古西部的地面风速明显增强。对比图 1 塔克拉玛干沙漠的情形，可以看出由于南疆盆地腹地为广阔的沙漠，明显不具备使地面风速在此增强的条件。而地面风速的大小对沙尘暴强度具有直接的影响[11]，因此，蒙古、我国内蒙古西部峡谷地形导致的地面风速增强是沙尘暴在该地区迅速增强的一个重要原因。

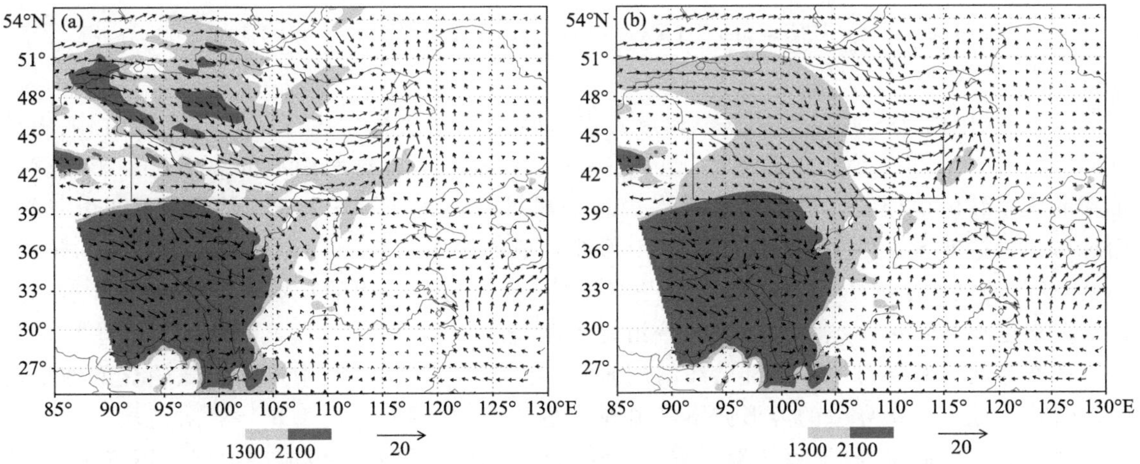

图 7　控制试验(a)和地形敏感试验(b)地形高度(阴影区，大于 1300 m)和 19 日 20 时地面 10 m 风场。

从图 7 还可看到，在阿尔泰—萨彦岭山地南坡为一致的西北气流(位于气旋冷锋后)，而在峡谷区为偏西气流，因此峡谷气流对这支下坡气流产生明显的阻挡。而图 5 中 19 日 14—20 时实况沙尘天气分布显示，此时在阿尔泰—萨彦岭山地南坡有大范围的沙尘暴产生(尽管山地上测站较少)，因此这支下坡气流是携带大量沙尘的，该气流被峡谷气流阻挡势必导致沙尘在峡谷区聚集，从而导致该地区沙尘暴的增强。从图 1 的统计结果可以看到，阿尔泰—萨彦岭山地南坡也是一个明显的沙尘天气多发区，与相同海拔高度的山地相比，这一特点非常明显。而山地的沙尘在下坡向南推进过程中被峡谷气流阻挡并聚集，显然是蒙古、我国内蒙古西部沙尘暴得以迅速增强的另一原因。

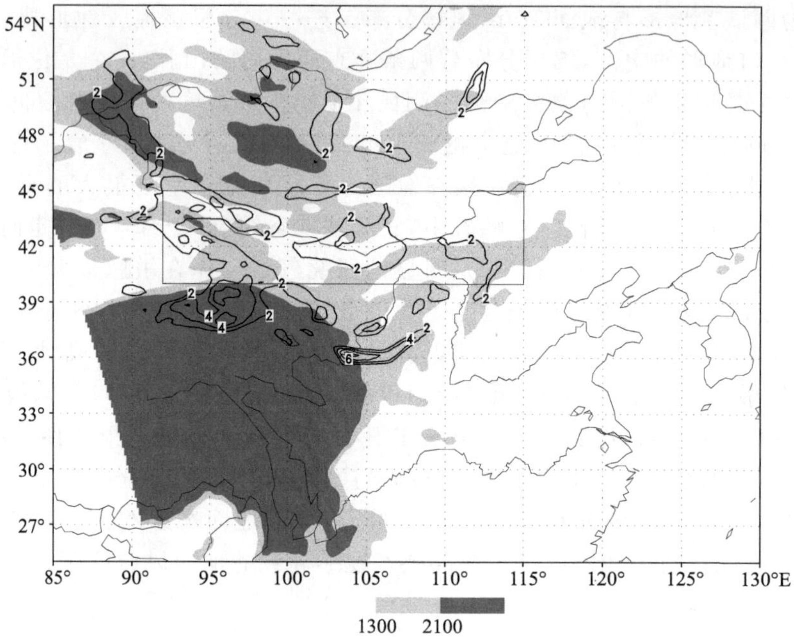

图 8 控制试验和地形敏感试验 19 日 20 时地面 10 m 风速差值(阴影区为控制试验地形高度,大于 1300 m)。单位:m/s 间隔:2

蒙古、我国内蒙古西部还有一个地形特点,即:与其他地区相比分布着较多的山脉、山地(图 2),这些山地虽然地形落差不大,但散布较广,数量众多。这些山地的一个重要影响就是地形起伏导致的地形波。这些地形波在山地迎风坡形成地形强迫抬升,背风坡则产生下沉。而地形强迫抬升无疑会加剧地面起沙。地形波生成的同时也自下而上地向对流层中的传播。因此,扬起的沙尘中有一部分将会被进一步上扬(参与这一过程的还有系统性的上升运动)。这一因素也将导致该地区沙尘暴增强。相对而言,较为平坦的南疆盆地显然不具备这样的条件。图 9 为两试验穿过蒙古南部的纬向剖面图,可以看到控制试验中地面有着明显的地形起伏。且这些地形起伏导致的地形波不仅出现在边界层中,同时也向上扩展到对流层。即使是非常微小的地形起伏,其导致的地形波也能够向上充分扩展,到达对流层中层。不仅于此,这些地形起伏也能够明显加大大气与地表的接触面积,从而有利于地面起沙。而敏感试验中在沙尘区上空基本上未出现波长较小的地形波。

为了衡量敏感试验地形改变之后对蒙古气旋的影响,对比了两试验气旋强度的变化,发现这一改变出乎意料的微小。19 日 20 时控制和地形试验气旋强度分别为 1006 hPa 和 1005 hPa,而 20 日 08 时两者均为 997 hPa。另外,通过图 6 风场得出的气旋中心位置来看,两者几乎完全一致。因此,可以说两试验沙尘强度的改变应该源自地形影响。地形因素是孤立于天气系统之外的自然因素,因而具有普遍性。需要进一步说明的是:蒙古、我国内蒙古西部沙尘暴“先增强,再扩展”的情形类似于在该地区大气中形成了可以称之为沙尘“中继站”的沙尘聚集区。它不同于塔克拉玛干沙漠沙尘扬起后即被吹走,在遇到高原阻挡才增强的情形。其意义在于:如果沙尘是首先在大气中实现高度聚集,然后进行扩展,则其扩展过程很大程度地遵循天气过程发展、演变规律,这样我们在追踪这些沙尘时就有章可循。

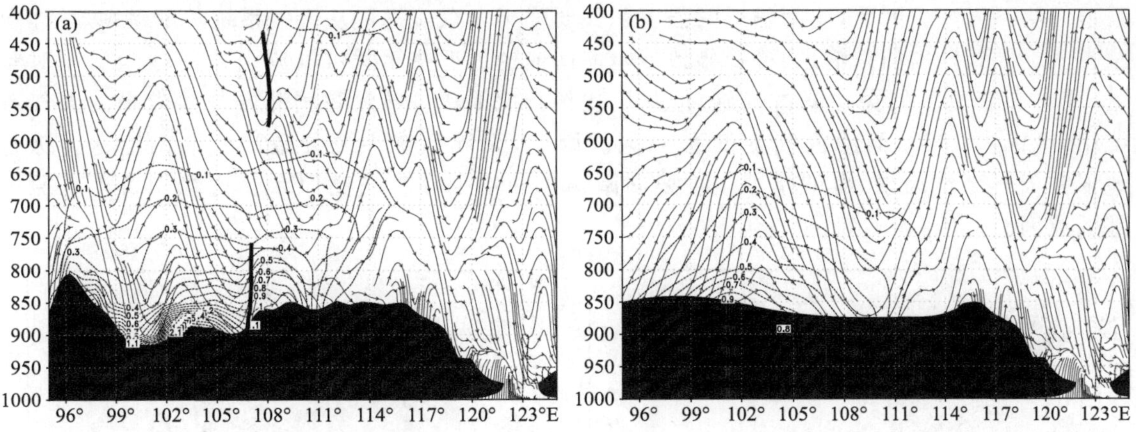

图 9 控制试验(a)和地形敏感试验(b)19 日 20 时沿北纬 42°纬向剖面图。等值线为沙尘浓度,单位 mg/m^3,间隔 0.1,流线利用 u;$w*50$ 合成

5.3 沙尘暴扩展阶段

沙尘暴在蒙古、我国内蒙古西部得到迅速加强后,开始向东和向南的扩展。此时仍然受到地形的影响。图 6c,d 为两试验 20 日 08 时地面沙尘浓度,可以看到由于前述的地形影响,沙尘强度已大为减弱。但撇开沙尘强度,从沙尘区扩展方向上可以看到,控制试验中沙尘主体向东扩展,而敏感试验中沙尘向南的扩展反而略强。表明地形的改变削弱了沙尘向东扩展的强度。

决定沙尘区扩展的核心因素是天气系统。对本次过程而言,该系统是蒙古气旋。由于沙尘区主要出现在气旋冷锋后部[7],因此,气旋后部气流方向决定了沙尘区的扩展方向。蒙古气旋从发展到成熟(或直至锢囚)的过程中,气旋冷锋后部气流方向基本上是从西北—东南向到准东—西方向最终变为西南—东北方向。因而沙尘区的扩展也经历从向南到向东直至向东北方向的改变。本阶段为气旋成熟阶段,因此,导致沙尘向东输送的一个因素是气旋冷锋后的偏西气流,地形改变不会影响这一气流方向。而导致沙尘向东输送的地形因素是前文提到的峡谷气流。由于萨彦岭山地南坡向南的下坡气流被峡谷气流阻挡,沙尘完成聚集之后部分沙尘将变为向东输送(图 7)。地形改变之后,向东的峡谷气流消失,沙尘向向东输送的强度随之减弱。图 10 两试验 u,v 差值能够验证这一点,图中在内蒙古中西部,敏感试验西风分量明显弱于控制试验,说明向东输送减弱;而在内蒙古中部,敏感试验北风分量明显强于控制试验,表明向南输送增强。可见,敏感试验沙尘主体向东扩展的减弱是由于地形因素。

图 1 中,我国沙尘天气的次多发区(也是沙尘源区之外发生沙尘天气最多的地区)位于青藏高原东北边缘到黄土高原中部一带。由于不是沙尘源区,因此这一多发区的形成显然是由于沙尘输送。解释这一现象需要综合分析我国北方沙尘输送的特点。如图 11,导致沙尘向南输送的地形强迫因素有两个,一个是前文得出的萨彦岭山地南坡的下坡气流,强度相对稍弱,另一个是青藏高原产生的地形强迫绕流,从图 7 可以直观地看到,由于青藏高原水平和垂直尺度很大,青藏高原东北边缘西北—东南方向的地形强迫绕流非常强大,从高原东北边缘一直到黄土高原中部,垂直方向可以达到对流层中层。而导致沙尘向东输送的地形因素仅仅是峡谷

强迫的峡谷气流。因此，向南输送的强度明显要强于向东输送。从天气系统角度考虑，总体上蒙古气旋导致沙尘向东输送，冷锋导致沙尘向南输送。近些年蒙古气旋引发的沙尘暴相对频繁，使研究者更多的关注蒙古气旋，但从历史统计来看[12]，导致我国北方强沙尘暴的天气系统中，冷锋占 76%，而蒙古气旋仅为 12%。因此，天气系统主要导致沙尘向南输送。综合地形和天气系统因素，也就是沙尘主要向南输送并集中在青藏高原东北边缘到黄土高原中部的原因。

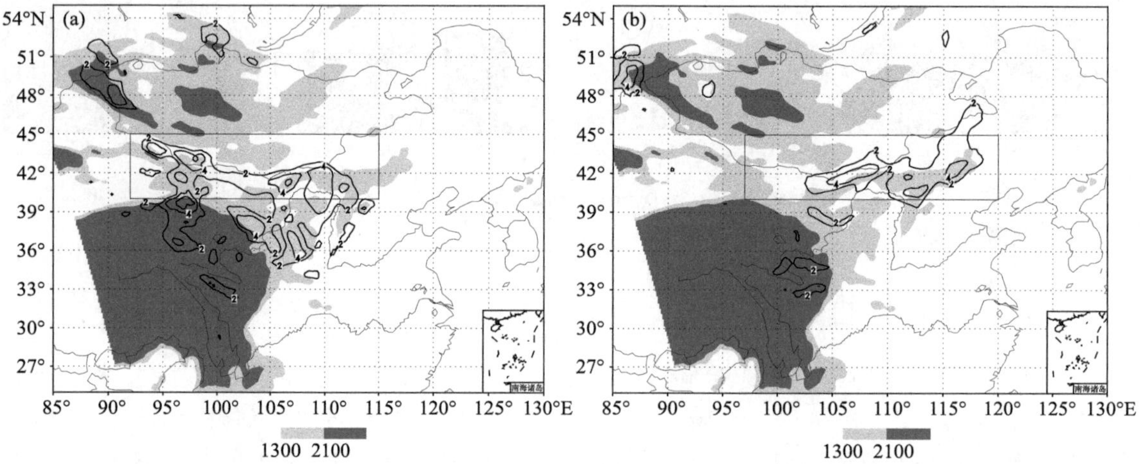

图 10　控制试验和地形敏感试验 20 日 08 时地面 10 m 风场 u 分量差值(a)和 v 分量差值(b)(阴影区为控制试验地形高度)。单位：m/s，间隔 2

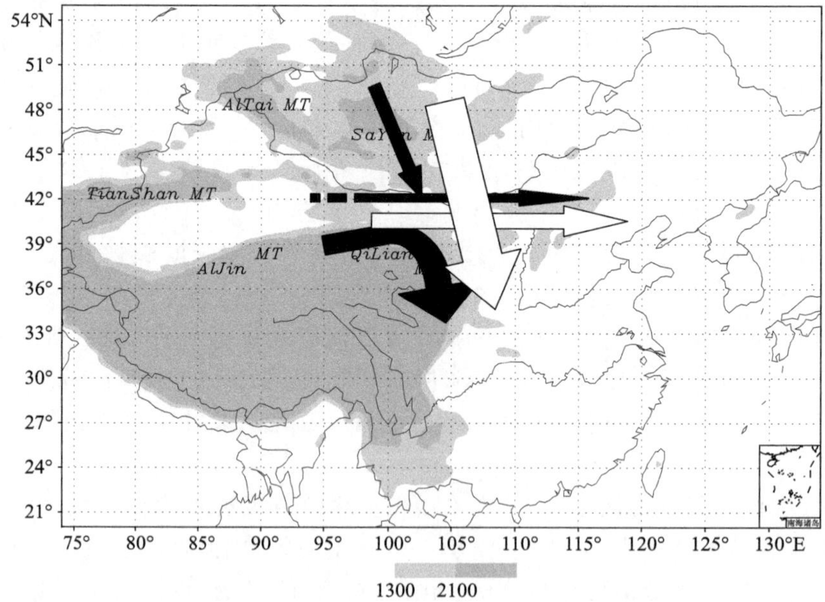

图 11　影响我国北方的沙尘输送示意图，其中，黑色箭头代表地形强迫因素(向东表示峡谷气流，向南表示下坡气流，高原东侧为地形绕流。白色箭头代表天气系统因素。向南代表冷锋，向东代表蒙古气旋)

6 讨论

本文基于观测资料探讨了东亚地形对沙尘输送的可能影响，同时，结合沙尘天气过程并利用数值模拟方法，初步研究了东亚地形对我国北方沙尘传输的影响。发现：东亚地区的两个主要的沙尘天气多发区一个位于我国南疆盆地，另一个位于我国内蒙古西部及蒙古南部。其中，南疆盆地的沙尘天气更多集中在盆地南缘；蒙古、我国内蒙古西部沙尘天气更多的集中在沙漠腹地。另外，青藏高原东北侧到黄土高原中部也是沙尘天气发生次数相对较多的地区。

沙尘天气多发区出现在蒙古、内蒙古西部沙漠腹地是由于这里具备了沙尘暴迅速增强的条件，首先是阿尔泰—萨彦岭南侧的峡谷地形强迫形成峡谷急流，明显增强了该区域地面风速。同时，萨彦岭山地南坡携带大量沙尘的向南下坡气流遇到东—西向的峡谷气流时受到阻挡形成沙尘的聚集。另外，这一地区起伏的地表产生的地形波加强了地面起沙。这种大量沙尘在大气中聚集之后再进行传输的特征可以视为在大气中形成了沙尘"中继站"。

导致沙尘从"中继站"进一步向东输送的地形因素是阿尔泰—萨彦岭山地南侧的峡谷地形，而萨彦岭山地南坡的下坡气流和青藏高原东北侧的地形强迫绕流是导致沙尘向南输送的原因。在天气系统角度，冷锋基本上导致沙尘向南输送，蒙古气旋成熟阶段冷锋后部偏西气流则产生向东的输送。由于青藏高原地形绕流的强大以及冷锋过程的相对频繁使得沙尘更多地向南输，这也是沙尘输送主要集中在青藏高原东北侧到黄土高原中部的原因。

参考文献

[1] 石广玉，赵思雄. 沙尘暴研究中的若干科学问题，大气科学，2003，**27**(4)：591-606.

[2] Zhang X Y，Gong S L，Zhao T L，et al. Sources of Asian dust and role of climate change versus desertification in Asian dust emission. *Geophys. Res. Lett.*，2003，**30**(24)：2272 doi：10.1029/2003GL018206.

[3] Zhao T L，Gong S L，Zhang X Y，et. al. A Simulated Climatology of Asian Dust Aerosol and Its Trans-Pacific Transport. Part I：Mean Climate and Validation. *J. Clima*. 2006，**19**(1)：88-103.

[4] Zhang X Y，Wang YG，Wang D，et. al. Characterization and sources of regional-scale transported carbonaceous and dust aerosols from different pathways in coastal and sandy land areas of China *Geophys. Res.*，2005，**110**：D15301，doi：10.1029/2004JD005457.

[5] Hacker J P，Mckendry I G and Stull R B. Modeled downward transport of a passive tracer over western north America during an Asian dust event in April 1998. *J. App. Meteor.*，2001，**40**(9)：1617-1628.

[6] 宋敏红，钱正安，蔡英，等. 中蒙强、弱沙尘暴年春季平均环流及沙尘活动变化分析. 气象学报，2007，**65**(1)：94-104.

[7] 姜学恭，沈建国，刘景涛，等. 导致一例强沙尘暴的若干天气因素的观测和模拟研究. 气象学报，2003，**61**(5)：606-620.

[8] 程丛兰，王迎春，刘伟东，等. 与动力场相耦合的区域沙尘分档模式及个例模拟. 科学通报，2004，**46**(19)：2007-2013.

[9] Anthes. R. A.，and T. T. Warner. Development of hydrodynamics models for air pollution and mesometeorolo-gical studies. *Mon. Wea. Rev*，1978，**106**(1)，1045-1078.

[10] 钱正安，蔡英，刘景涛，等. 中国北方沙尘暴研究若干问题. 沙尘暴学术研讨会文集，2002，台湾"蒙藏委员会"出版：63-76.

[11] 王小玲,翟盘茂.中国春季沙尘天气频数的时空变化及其与地面风压场的关系.气象学报,2004,**62**(1):96-103.

[12] 刘景涛,钱正安,姜学恭,等.中国北方特强沙尘暴的天气系统分型研究.高原气象,2004,**23**(8):540-548.

An observational and numerical Study on the topography influence on the dust transportation

JIANG Xuegong[1] CHEN Shoujun[2]

1. Inner Mongolia Autonomous Regional Meteorological Observatory, Huhhot, 010051;
2. Department of Atmospheric Science, the college of physics, Peking University, Beijing 100871

Abstract Through the observational statistic analysis and numerical simulation, a study on the influence of topography in eastern Asia on the dust transportation was conducted. The result showed that the high frequency dusty weather regions locate at the west of Inner Mongolia and southwestern Mongolia, especially at the hinterland of deserts, and also locate at the southern Xinjiang basin, especially at the south of it. The second high frequency dusty weather region is on the northeast of Qinghai-Tibet plateau. The dusty weather more frequently occurred at western Inner Mongolia and southwestern Mongolia is owing to some factors favorable for the dust storm formation besides plenty of dust particles can be provided there. Firstly, the canyon jet, which forms at south of Altai-Sayan mountain, enhanced the surface wind speed and so in favor of the buildup of dust storm. Secondly, the southwardly down slope stream carrying a mass of dust particles from Sayan mountain was obstructed by the eastwardly canyon jet and so the dust particles concentrate there and dust storm boosts up. Thirdly, most mountains and hilly lands exist there, which produce the topographical wave and boost up the dust rising. The pattern that dust particles assemble in the atmosphere (form a severe dust storm) before they transport can be say a "dust relay station" forming there.

The topography also affects the spreading of dust storm. The eastwardly canyon jet enhances the eastwardly transportation of dust particles from the "dust relay station". The down slope stream from the Sayan mountain and the topography forcing circum-stream at the northeast of the Qinghai-Tibet plateau enhances the southwardly transportation of dust particles. Owing to the circum-stream is so strong and the occurrence of cold front is so frequent, which transport the dust particles southwardly too, that the dust transportation in northern China is mostly southwardly and concentrates at west of Huangtu plateau.

Key words: dust storm, topography, observational and numerical study

基于 CALIPSO 资料的沙尘暴过程沙尘垂直结构特征分析*

姜学恭[1] 陈受钧[2] 云静波[1]

1. 内蒙古自治区气象台，呼和浩特 010051；2. 北京大学物理学院大气科学系，北京 100871

摘 要：综合利用 CALIPSO 星载激光雷达探测资料和 MM5 数值模拟结果，对 2010 年 3 月 19—22 日强沙尘暴过程不同阶段沙尘垂直分布特征及其动力、热力结构进行了初步分析，结果发现：在沙尘暴成熟阶段，沙尘层分布于 2～9 km(850～250 hPa)的几乎整个对流层中，冷锋前抬升和锋后下沉导致的旺盛垂直混合使沙尘呈现相对均匀的垂直分布。在沙尘扩展及远距离传输阶段，沙尘层明显分为两层，分别位于对流层低层(700 hPa 以下)和对流层中高层(600～300 hPa)。在沙尘暴各个阶段，弱风速垂直切变和弱位温、相当位温垂直变化始终与沙尘层配合，显示沙尘层维持中性混合层，而两个沙尘层之间则为强风速垂直切变及位温、相当位温锋区。另外，沙尘暴发展过程中，高空急流、位涡、比湿等要素均表明出现明显的对流顶折叠和高空位涡下传，且在对流顶较高的区域，沙尘向上的扩展也较高，反之则较低。需要指出，在沙尘暴扩展和远距离传输阶段，在 40°N附近，7～9 km 沿纬向一线，均出现一小范围孤立沙尘区位于平流层中(或平流层附近)，表明沙尘暴过程中能够产生沙尘的对流层—平流层输送，并在平流层中形成持续性的沙尘传输带。这可以成为沙尘气溶胶对流层—平流层输送及其在平流层中传输的一个直接的监测证据。

关键词：沙尘暴，沙尘，垂直结构，CALIPSO 资料

1 引言

东亚沙尘暴形成于东亚和我国北方沙漠和荒漠化地区，通过沙尘远距离传输对我国大部地区、东亚、西太平洋等地产生严重影响。近年来，研究者对沙尘暴的形成和影响开展了广泛的研究，王伏村[1]等对河西走廊一次特强沙尘暴过程进行了天气学和诊断分析，揭示了地面热通量、动量通量、锋生作用等对沙尘暴发生发展的影响。王丽娟、赵琳娜等[2]对 2009 年一次强沙尘暴过程的影响系统、沙尘气溶胶浓度特征进行了分析，并研究了沙尘暴过程起沙、远距离传输等特征。郭本军等[3]利用中韩沙尘暴合作项目的脉冲激光雷达探测资料对 2008 年一次沙尘暴过程中大连市的沙尘气溶胶垂直分布特征进行了初步分析。郭铌等[4]则研究探讨了利用 MODIS 资料进行沙尘暴判识的方法。这些研究对揭示沙尘暴形成机理、指导沙尘暴预报及进行沙尘暴影响的评估等具有一定的指导意义。

沙尘暴过程不仅是大气运动和演变的过程，也是大气与沙尘粒子相互作用的过程。因此，研究沙尘暴发生发展机理及其影响，需要全面地掌握沙尘暴过程中沙尘的运动和演变状态。目前，尽管各类天气监测手段能够对大气运动过程进行较详尽的刻画，但对于沙尘暴过程中沙尘的运动和演变却难以进行全面的揭示。例如，沙尘天气的地面观测只能给出地面水平能见

* 资助项目：国家自然科学基金项目(41365003，40965005)。

度，沙尘气溶胶观测只能给出地面附近沙尘粒子的浓度、大小、类型等特征，卫星遥感则只能给出沙尘层顶部的特征，或整个沙尘层的光学厚度，但对于大气中沙尘的垂直分布特征却难以描绘。这也成为制约深入揭示沙尘暴形成、传输机理的重要因素。

云—气溶胶激光雷达和红外探路者卫星观测(Cloud-Aerosol Lidar and Infrared Pathfinder Satellite Observations，CALIPSO)是由 NOAA 为主开展的结合了激光雷达探测器和红外、可见光成像器的综合卫星观测，能够对全球范围内云、气溶胶的垂直结构和特征进行探测。利用该资料提供的沙尘暴过程中不同阶段的沙尘垂直分布特征，无疑有助于深入揭示沙尘暴发生、演变特征和规律。为此，本文拟针对 2010 年 3 月 19—22 日我国北方一次强沙尘暴过程，利用 CALIPSO 资料，并结合中尺度模式 MM5 模拟结果，对沙尘暴不同阶段大气中沙尘垂直分布特征及其动力、热力结构特征进行分析，以期更深入地揭示沙尘暴的发生规律。

2　天气过程概述

2010 年 3 月 19—20 日强沙尘暴过程是一次典型的冷锋型沙尘暴过程。其影响地区包括西北地区大部、内蒙古中西部、华北及东北地区，并在内蒙古达到最强。

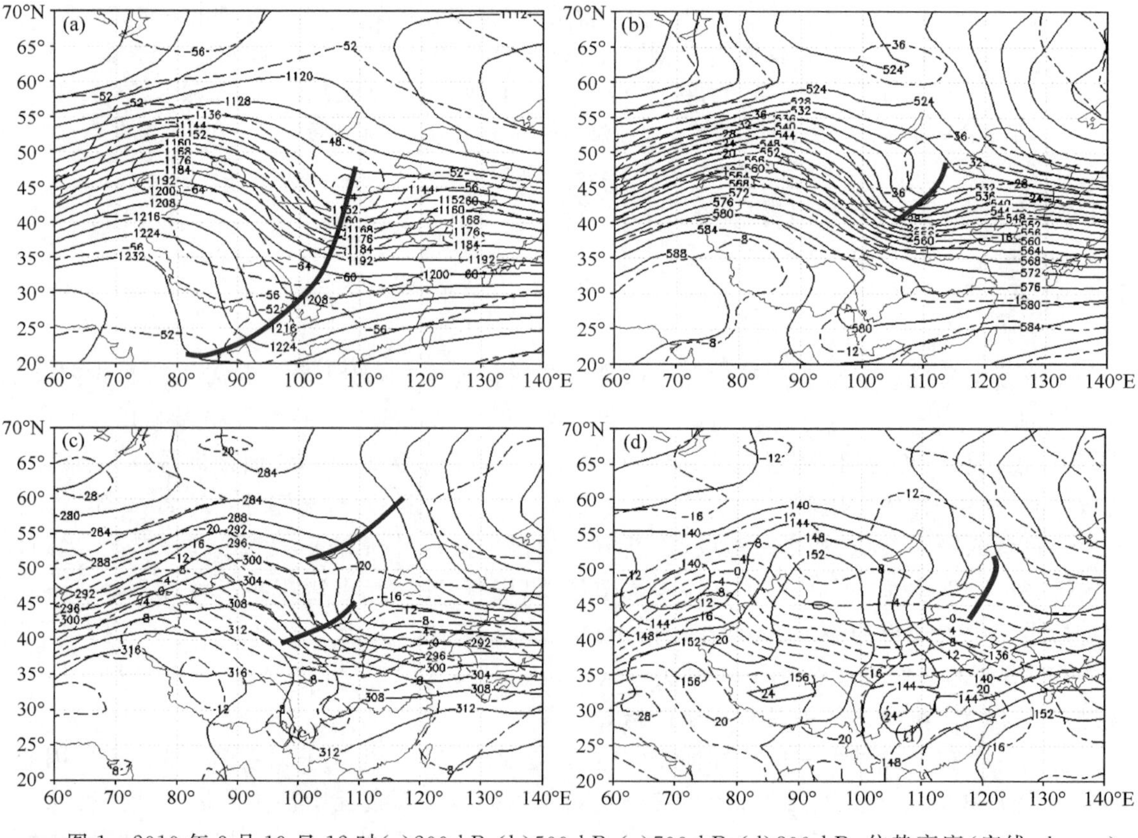

图 1　2010 年 3 月 19 日 12 时(a)200 hPa(b)500 hPa(c)700 hPa(d)800 hPa 位势高度(实线，dagpm)和温度(虚线，℃)。粗实线为槽线

从形势场演变来看,3 月 18 日 12 时(世界时,下同),欧亚大陆维持两槽一脊,欧洲大陆与太平洋西岸为大槽,乌拉尔山地区为高脊,我国北方大部地区受脊前西北气流控制。地面图上,冷高压位于阿尔泰山以西,萨彦岭山地背风坡则有低压形成。19 日 00 时,乌拉尔山高脊向北加强,导致槽前西北气流增强,同时,700 hPa 在蒙古国中部形成东北—西南向横槽,引导冷空气向东南方向移动。地面冷锋越过萨彦岭山地,推进至内蒙古西部,并在蒙古南部、内蒙古西部、甘肃北部引发沙尘暴。19 日 12 时,高空横槽自西北向东南方向南压,700 hPa 锋区推进至内蒙古中部,等温线和等高线接近垂直,冷平流强盛。地面图上,随着冷高压南压,地面大风大幅增强,平均风力达 20 m/s,沙尘暴天气范围明显增大,海力素出现黑风暴,本次过程沙尘暴也达到最强。20 日 00 时,高空横槽加强并转竖,向东偏南方向移动,推进至东北—华北一线,槽区强冷平流继续维持。地面冷锋则进一步向东偏南方向推进,扬沙和浮尘天气扩展至山西、河北、山东等地。20 日 12 时,高空槽移至朝鲜半岛,并完全转竖,同时,温度槽逼近高度槽,斜压性逐渐减弱。地面图上,冷锋明显减弱,冷高压南压趋势增强,受高空气流引导,地面低压转向北偏东方向快速移动。沙尘(浮尘)天气则出现在山东、东海及朝鲜半岛等地。21 日 00 时,高度槽闭合形成低涡,斜压性进一步削弱。地面冷高压明显减弱,浮尘天气则扩展至东海东部、日本岛等地。

3 模式和方案

模拟采用中尺模式 MM5。MM5 多用于对对流层天气过程的预报和模拟,但国内外一些研究也利用 MM5 对对流层—平流层输送进行模拟研究,如 Ravetta 等[5]利用 MM5 成功的模拟了一次对流层顶折叠事件,模拟的对流顶附近的风场、温度场与观测结果较为吻合。杨健等[6]利用 MM5 对 2000 年春季一次切断低压引发的对流层—平流层交换过程进行了成功的模拟研究。

模拟采用两重网格嵌套,粗网格格点数 110×90,中心(114°N,53°E),格距 45 km,细网格格点数 193×112,格距 15 km,模式层顶 200 hPa,垂直方向为 32σ 层。采用 Blackadar 高分辨率行星边界层方案和云辐射方案。模式背景场为每日 4 次的 2.5°×2.5°N CEP 再分析资料,模拟时段为 2010 年 3 月 19 日 00 时—22 日 00 时,模拟时间 72 小时。

图 2 为 2010 年 3 月 19 日 18 时 NCEP 再分析(实线)及 MM5 模拟(虚线)500 hPa 高度

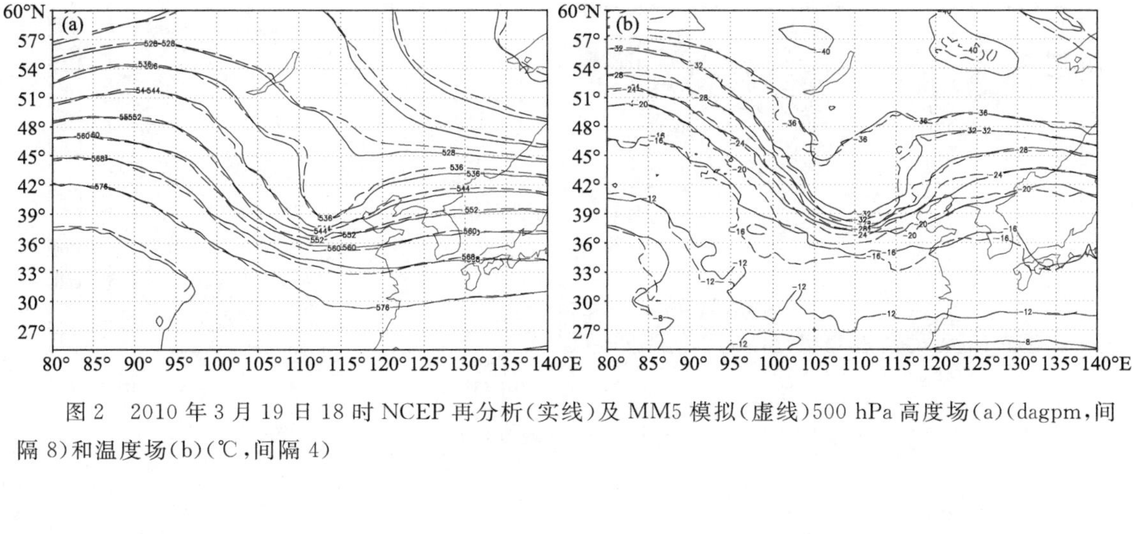

图 2 2010 年 3 月 19 日 18 时 NCEP 再分析(实线)及 MM5 模拟(虚线)500 hPa 高度场(a)(dagpm,间隔 8)和温度场(b)(℃,间隔 4)

(a)和温度(b),可以看到,模拟的 500 hPa 低槽位置、强度等于实况较为吻合,特别是 39°N 附近的低槽。模拟的 500 hPa 温度槽的位置、强度也与 NCEP 分析场也较为一致。只是在 57°N、125°E 及 36°N、98°E 等外围区域出现一定的模拟偏差。总的来说,模式成功模拟了天气系统的发展演变过程。

4　结果分析

4.1　沙尘暴成熟阶段

携带激光雷达的 CALIPSO 卫星沿极地轨道运行,即沿经圈过南北两极的轨道对大气进行监测,因此,雷达监测的时间和区域与本次沙尘暴过程所关注的时间和区域并不完全一致。为此,考虑监测的区域并结合本次沙尘暴发生和演变过程,选择 2010 年 3 月 19 日 19 时、20 日 18 时、21 日 18 时三个时间点雷达监测资料,分别代表沙尘暴不同阶段进行分析。

图 3a 为 CALIPSO 的星载激光雷达监测的大气气溶胶类型及其垂直分布,图 3b 为卫星运行轨迹(http://www-calipso. larc. nasa. gov/),为了更清楚地看到沙尘垂直分布,图 3c 为图 3a 的局部放大图(下同,并略去原图)。可以看到,在 31°—37°N,103. 8°—105. 5°E 间,沙

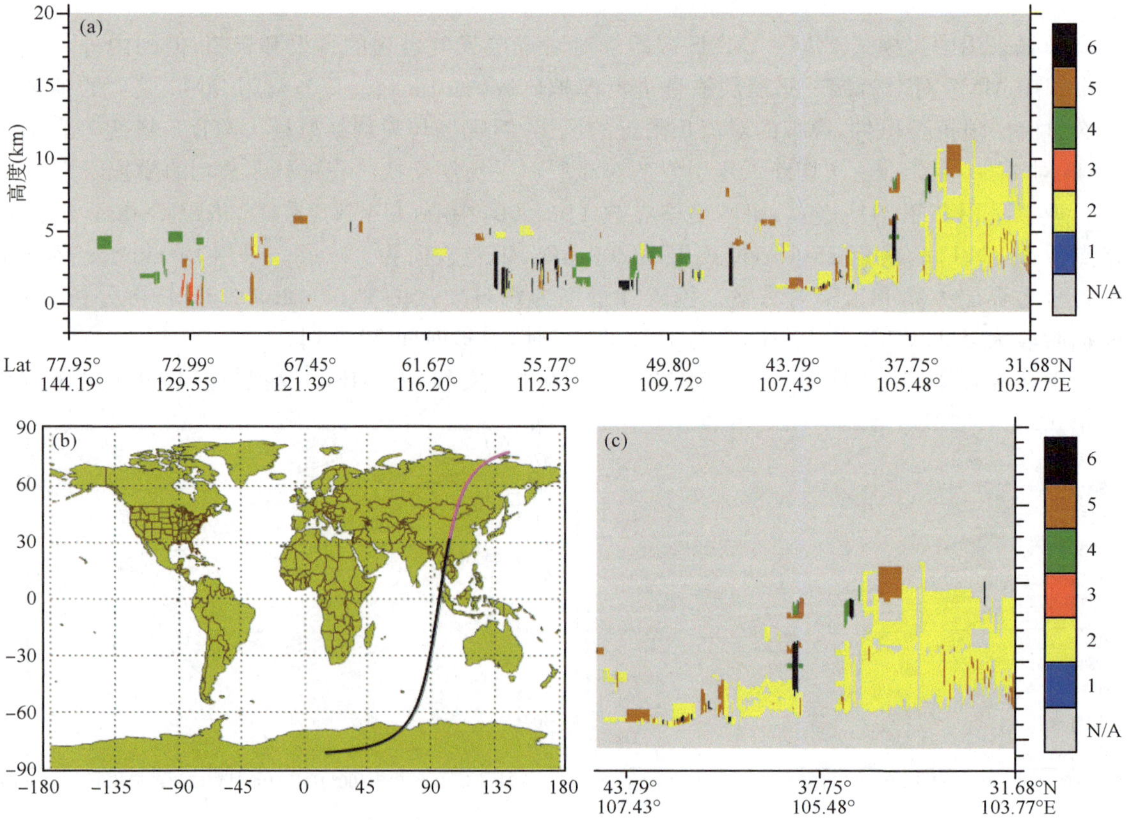

图 3　2010 年 3 月 19 日 19 时 CALIPSO 监测的气溶胶分类和垂直分布及卫星运行轨迹

尘区位于 2～9.5 km 之间，在 37°N 以北，37°—40°N，105.5°—106.5°E 范围内，沙尘区位于地面 2～4 km 的对流层低层，另外，在 39.3°N，6～7 km 附近有一小范围孤立沙尘区存在。

前文分析表明，本次沙尘暴过程于 3 月 19 日 12 时前后发展成熟并达到最强，19 日 18 时，沙尘暴略向东南方向扩展，但强度较之前 6 小时变化不大，因此，19 日 19 时大体上可以代表沙尘暴成熟阶段。从图 4 地面图及雷达监测带位置来看，地面低压、高压连线为东南—西北向，监测带大致位于地面冷锋后部。高空则位于冷槽后部，由于锋区位于 30°—40°N 之间，呈东西方向，因此，监测带大致垂直穿过高空锋区。

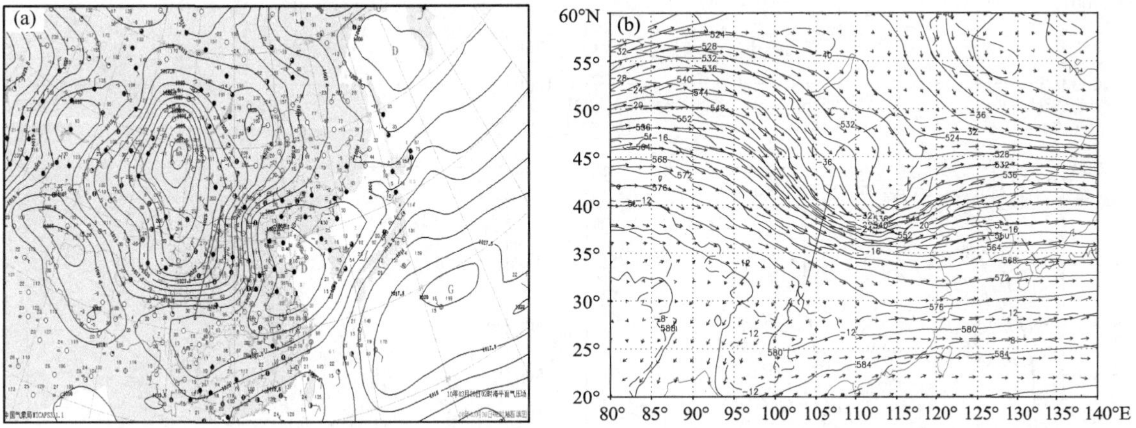

图 4　2010 年 3 月 19 日 18 时地面天气图(a)及 500 hPa 高度场和水平流场(b)，图中实线为雷达监测带

图 5 为沿图 3 雷达监测带的垂直剖面图，首先，结合图 4，可以看到，地面冷锋南支大致位于 31°N 附近，因此，图 5a 中 31°—34°N 间上升区位于冷锋锋前，下沉区位于冷锋后；而 35°—37°N 附近明显有副冷锋存在。对比位势高度，可以看到 31°—37°N，103.8°—105.5°E，2～9.5 km 范围内的沙尘区，对应于 700～300 hPa 的整个对流层，锋前、锋后(副冷锋前、锋后)，即上升运动区和下沉运动区沙尘浓度差异并不明显。这可能由于冷锋前后上升、下沉运动导致的旺盛垂直混合所致。

另外，在 37°N 以北，37°—40°N，105.5°—106.5°E，2～4 km 高度的沙尘区，对应于 600 hPa 以下，700 hPa 为中心的对流层低层。结合图 4 可以看到，该区域位于地面高压东侧，冷锋为自西偏北向东偏南推进，因此风速明显强于前述向南扩展的一支。向东扩展的沙尘主要位于 700 hPa，而向南扩展的沙尘则散步于几乎在整个对流层。其可能形成原因是：冷锋东支所处的 37°—43°N 间为西风急流区，由于与强盛西风气流配合，沙尘尚未形成进一步上扬即被西风气流吹走，因此向下游扩展明显，但沙尘垂直上扬不高；而与南支冷锋配合的西风气流较弱，因此沙尘在锋面抬升作用下产生明显的扬升。

大气位涡源位于极地平流层中，中高纬度低槽、低涡等天气系统的发展加深，伴随高空急流的下落和对流层顶的折叠，常常形成极地平流层高位涡向南、向下的传输，相关研究将位涡为 2 pvu 定义为对流层顶[7]。从图 5b 位涡垂直分布来看，在沙尘主要分布区的 37.5°N 以南，位涡均小于 1 pvu。表明沙尘位于对流层中；而在 37.5°N 以北，对流顶出现了明显的下落，在 37.8°—39.8°N 间下落至 500 hPa(5500 gpm 左右)，并与自北向南的倾斜下沉气流伴随。可见，在对流顶较高的区域，上升运动和垂直混合扩展的高度较高，沙尘区出现的高度也较高

(9.5 km,300 hPa)。而在对流顶明显下落的区域,主体沙尘的高度较低(4 km,600 hPa)。

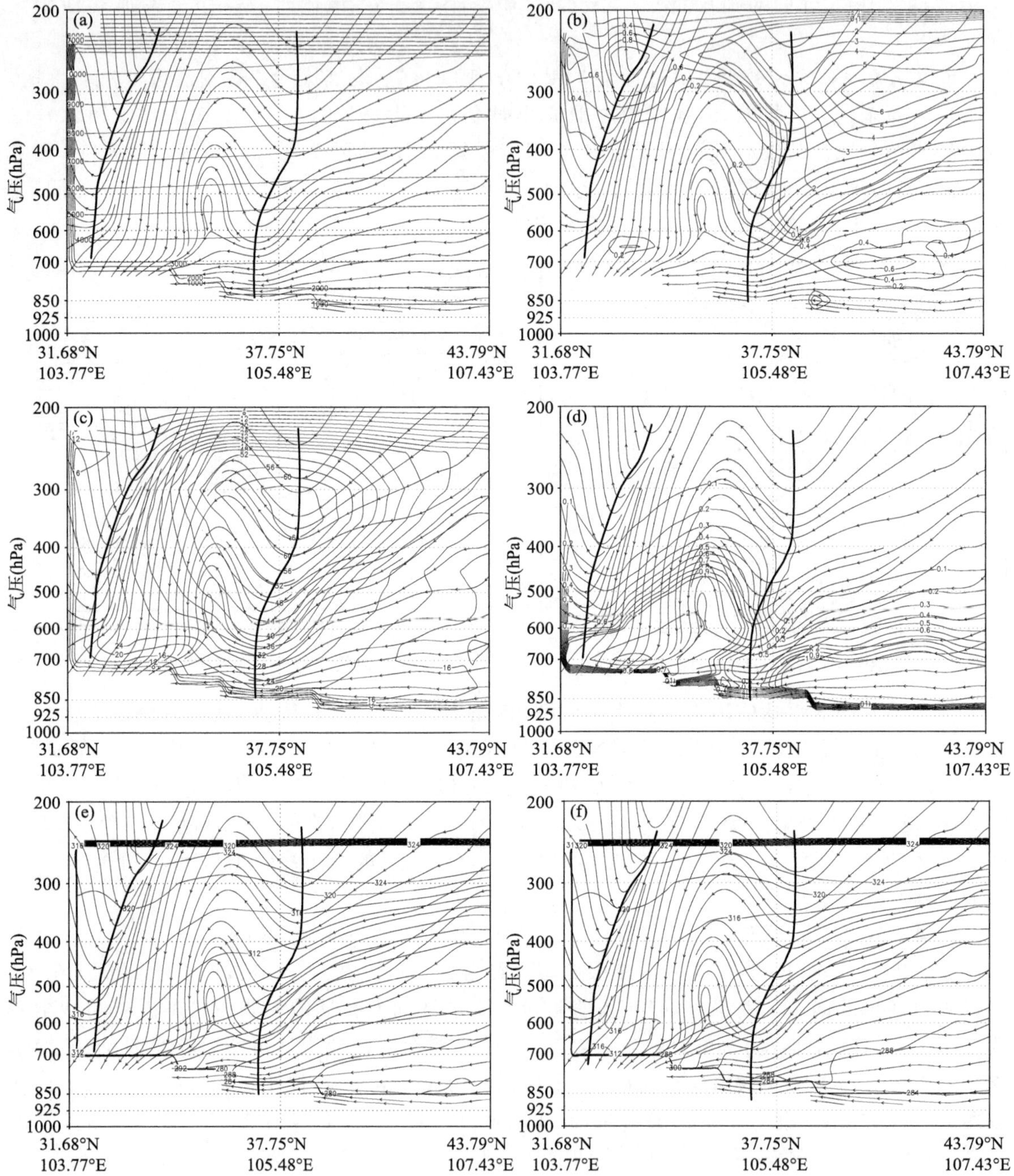

图 5　MM5 模拟的 2010 年 3 月 19 日 19 时沿雷达监测带的垂直剖面图,其中,流场为沿剖面水平风与 $W*100$ 合成,等值线分别为(a)位势高度,gpm;(b)位涡,pvu;(c)全风速,m/s;(d)混合比,g/kg;(e)位温,K;(f)相当位温,K

需要指出,前文提到的 39.3°N 附近,6～7 km 的微小孤立沙尘区,图 5b 显示该位置对流顶已下落至 500 hPa(5500 gpm 左右),可见,该沙尘区位于平流层中,这表明之前曾经发生了沙尘气溶胶的对流层—平流层输送。对于沙尘气溶胶的对流层—平流层输送问题,目前国内外开展的研究尚不是很多,Li 等[8]利用 SAGE Ⅱ 气溶胶观测资料通过气候学统计对比,初步验证了东亚沙尘气溶胶在平流层中传输的现象。作者等利用数值模拟方法对东亚沙尘气溶胶的对流层—平流层输送问题开展了一些探索性分析研究[9],但总的来说,对于沙尘气溶胶对流层—平流层输送及沙尘在平流层中的传输还缺乏基于天气监测的直接证据。而上述基于雷达监测的沙尘区位于平流层中的现象,说明本次沙尘暴过程中确实有沙尘在平流层中进行传输,这无疑是沙尘气溶胶对流层—平流层输送及其在平流层中传输的一个直接监测证据。考虑到该剖面图为准经向剖面,垂直于剖面的主体气流为自画面流出的偏西气流,因此该孤立沙尘区表明有一自西向东的沙尘传输带位于平流层低层(450 hPa)。

全风速和比湿的垂直分布可以进一步说明对流顶的下落。首先,从图 5c 全风速垂直分布可以看到,在 37.5°N 以北,高空急流出现了明显的下落,64 m/s 的急流核位于 350 hPa(常态大气急流位于 250 hPa 左右)。从其位置来看,该区域较图 5b 中 2 pvu 高位涡的下落区略偏南,表明急流下落的动力作用是对流顶折叠和高空位涡下传的驱动因素[9]。从比湿分布可以看到(图 5d),与 2 pvu 的对流顶下落区相对应,对流顶下落区的比湿均较小,在 PV＞1 pvu 区域,比湿基本上小于 0.1 g/kg,表明该区域为源自平流层的干燥空气,这符合平流层大气的基本特征(一些研究利用比湿大小界定对流层和平流层[10])。

另外,从图 5c 还可看到,在 37.5°N 以南较均匀地散布于整个对流层中的沙尘区(4～9 km,300～600 hPa),以及 37.5°N 以北 600 hPa 以下的沙尘区,风速垂直切变均很小。图 5e、f 位温、相当位温垂直分布则显示,37.5°N 以南 300～600 hPa,及 37.5°N 以北 600 hPa 以下的沙尘区,位温、相当位温垂直变化小于 1～1.5 K/km。(相关研究将位温垂直变化小于 1.5 K/km 界定为中性层结[14])。一些研究表明[11～13],在沙尘暴形成及沙尘远距离传输过程中,沙尘层往往与中性混合层相伴。这一方面源于沙尘暴发生时(冷锋触发)的旺盛垂直混合,另一方面,沙尘气溶胶的辐射强迫则是另一原因。沙尘气溶胶的存在,一方面阻挡地面接收太阳短波辐射,其自身又释放长波辐射加热大气,从而削弱了常态大气白天由于地面加热导致的不稳定趋势(夜间则相反,削弱地面冷却导致的稳定趋势),使大气层结趋向中性[13]。本次过程中,沙尘层的层结呈现中性特征,而其弱风速垂直变化,更直观的反映了大气旺盛垂直混合的状态特征。可见,上述监测事实可以成为沙尘层与中性混合层相伴的另一直接证据。

4.2　沙尘暴扩展阶段

24 小时后,3 月 20 日 18 时(图 6a),锋后高压已沿西北—东南向推进至华东地区(中心 33°N),锋前低压则沿西南—东北向移至日本北部(中心 42°N,130°E),地面沙尘区已扩展至黄海以东地区。高空槽移至朝鲜半岛以东,日本岛以西(图 6b)。雷达监测带位于地面低压后部,高空则位于槽后,横穿槽后西北气流区(与高空槽的位置关系和上节情形相近,但略滞后)。可见,雷达监测大体上横穿沙尘向东的输送带。

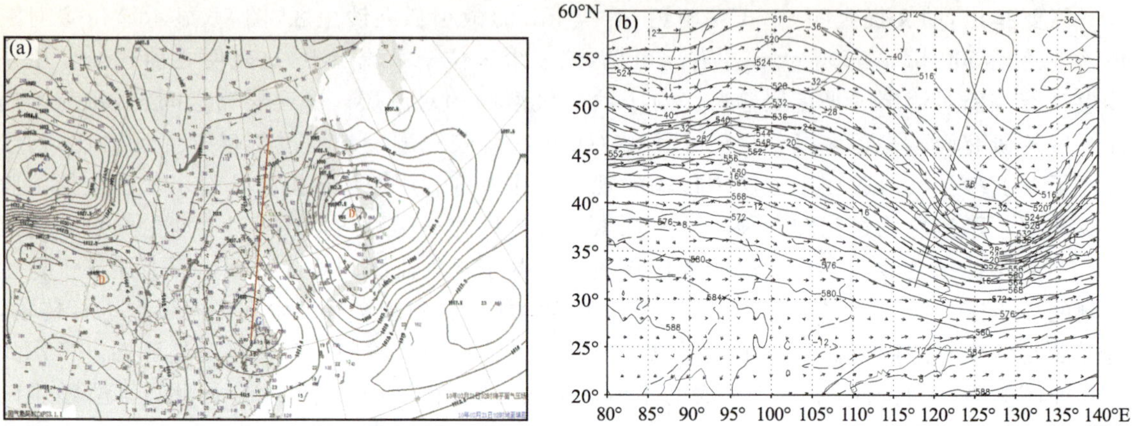

图 6　类似图 4，时间为 2010 年 3 月 20 日 18 时

从沙尘垂直分布来看(图 7)，与之前相似的是沙尘主体依然分布在 31.3°—40°N 一带，并向东推进，与之前不同的是沙尘层明显分为 2 层，一层位于 2.5 km 到地面的对流层低层，一层位于 5～10 km 的对流层中高层，之间为无沙尘区。同时，同样在 40.3°N，120°E 附近 8～9 km 存在孤立的沙尘区。可见，沙尘垂直分布体现了沙尘传输的连续性及变化。

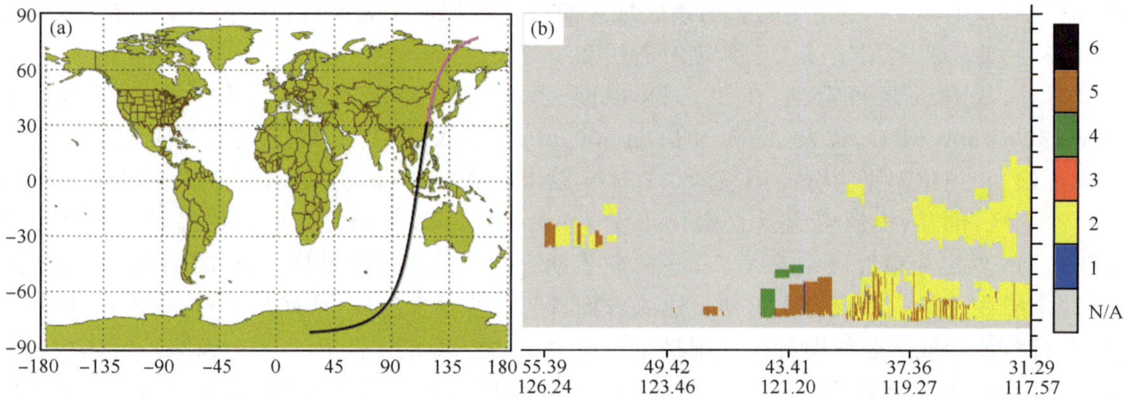

图 7　类似图 3，时间为 2010 年 3 月 20 日 18 时

从图 8a 剖面图流场可以看到，40°N 以南为自北向南的下沉气流，与图 7 沙尘区配合。另外，32°N 附近存在一明显的垂直环流，南侧为上升支。对比图 7 沙尘分布可以看到，32°N 以南与垂直环流上升支对应的区域，沙尘区分布于 7～10 km，而其北侧为 5～7.5 km。可见，受垂直环流上升支的驱动，沙尘层高度明显高于其北侧。

从图 8b 位涡来看，以 37°N 为界，其南侧无明显位涡下传，其北侧则形成明显高位涡下传。与偏北下沉气流配合，代表对流顶的 2 pvu 下落到 500 hPa 以下，位于 38°—41°N 一带。而沙尘垂直分布显示 37°N 以南为沙尘主要分布区，北侧则较少。可见，与上节讨论的类似，在对流顶较高的区域，沙尘区向上扩展也越高。另外，在 40.3°N，120.2°E，8～9 km 为孤立沙尘区，范围较之前 24 小时有所扩大，位涡剖面图显示该位置位涡值为 5 pvu(2 pvu 已下落至 500 hPa 以下，对应高度 5.5 km(图 8a))。可见，该沙尘区已深入平流层中。从沙尘传输的持续性来看，其显然与之前 39.3°N 的孤立沙尘区来自同一沙尘传输带，且范围增大。

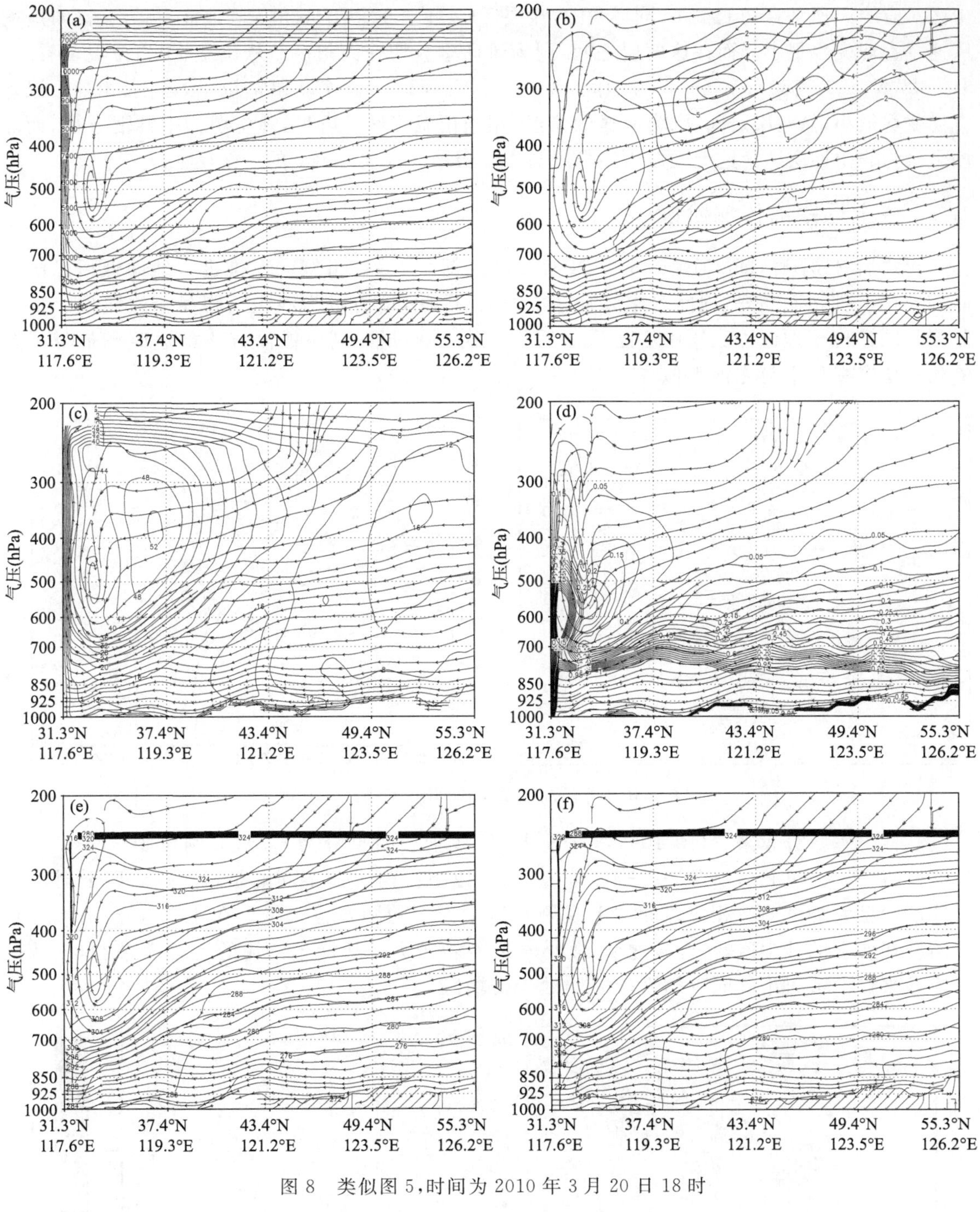

图8 类似图5,时间为2010年3月20日18时

全风速和比湿垂直分布同样验证了对流顶的下落。图8c中,52 m/s高空急流核已下落至400 hPa,且(37°N)同样位于2 pvu高位涡下落区的南侧,与上节讨论的情况一致。从比湿来看(图8d),沿偏北下沉气流,0.1 g/kg等值线已向南向下扩展至37°N,600 hPa以下,与2 pvu区对应的比湿均维持在0.05 g/kg以下。可见,全风速和比湿进一步验证了对流顶下落的事实。

从风速、位温、相当位温分布来看。图8c中,在31°—37°N,550~250 hPa(5~10 km)间,与沙尘区对应的风速垂直切变很小，同样，在对流层低层 31°—40°N，750 hPa 到地面，与沙尘区对应的风速垂直切变也很小。而在这两层之间，则为风速垂直切变大值区。从位温、相当位温来看，在 31°—37°N，550～250 hPa(5～10 km)间，及 31°—40°N，2.5 km(750 hPa)以下，位温、相当位温垂直切变均很小，而两层之间，则为明显的位温、相当位温锋区。可见，沙尘层为中性混合层，而非沙尘区为非中性层结。这无疑成为沙尘层与中性层结配合这一事实的一个反证。

4.3　沙尘远距离传输阶段

再之后 24 小时，即 2010 年 3 月 21 日 18 时(图 9)，系统继续向东推进，，地面低压已沿东北方向移至鄂霍次克海，地面沙尘区则基本消失，因此，该阶段为沙尘远距离传输阶段。雷达监测带同样位于地面冷高压中心至低压后部一线，500 hPa 高空槽后方。雷达监测带与系统的相对位置总体上与之前类似。

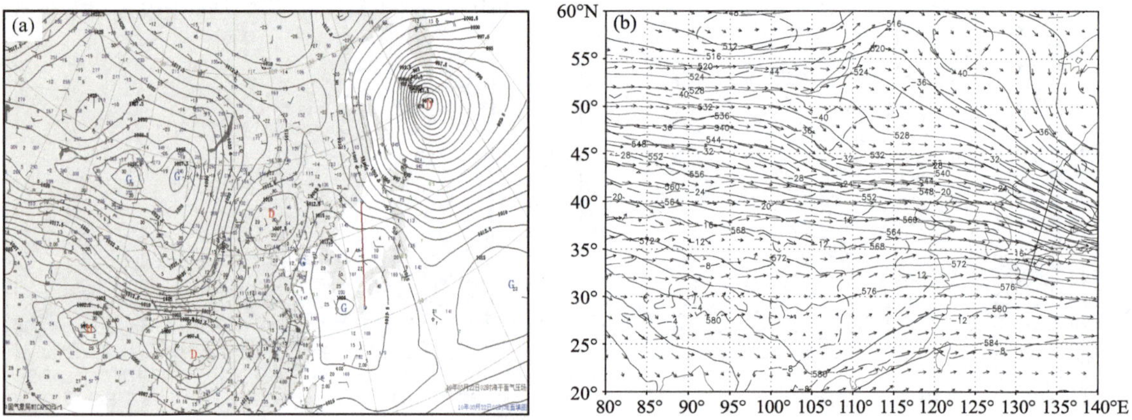

图 9　类似图 6，时间为 2010 年 3 月 21 日 18 时

从沙尘垂直分布来看(图 10)，与之前类似，沙尘层明显分为两层，分别位于 31°—37°N，130°—133°E，4～9 km，以及 30°—40°N，1.5 km 以下。与之前不同的是，1.5 km 以下对流层低层沙尘已大为减弱。相应地面图上观测到沙尘天气的测站也很少。在 40°N，8～9 km 间，依然存在孤立的沙尘区，范围较之前 24 小时进一步扩大。从沙尘传输的持续性来看，该沙尘区应该来自上节讨论的同一纬度(40°N)沙尘传输区。

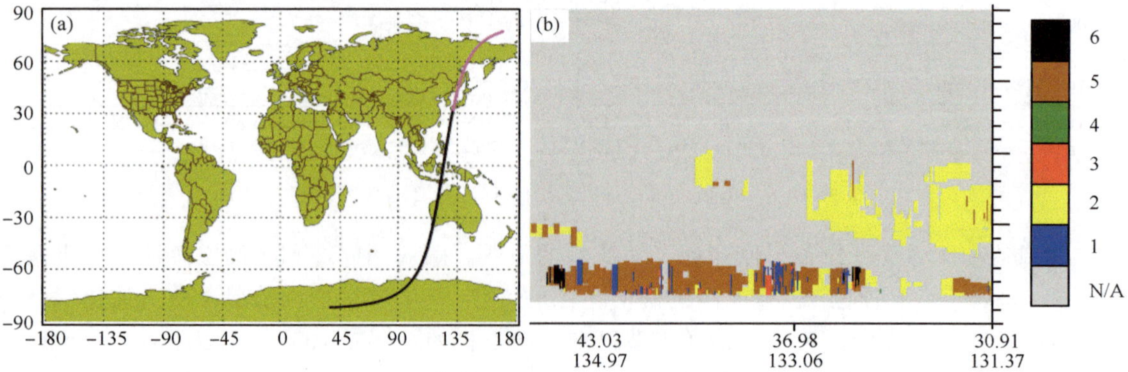

图 10　类似图 7，时间为 2010 年 3 月 21 日 17 时

从剖面流场来看(图 11),雷达监测带总体仍为偏北下沉气流,且强度较之前明显偏强。与沙尘区配合同样为下沉运动区。从位涡分布来看,43°N 附近 350 hPa 以上存在 2 pvu 的位涡高值区。对于 40°N,8～10 km 附近的孤立沙尘区,其大约位于 350～250 hPa 间(图 11a),对应的位涡值约为 1 pvu 左右。因此,该区域接近但并不在平流层中。考虑到该孤立沙尘区

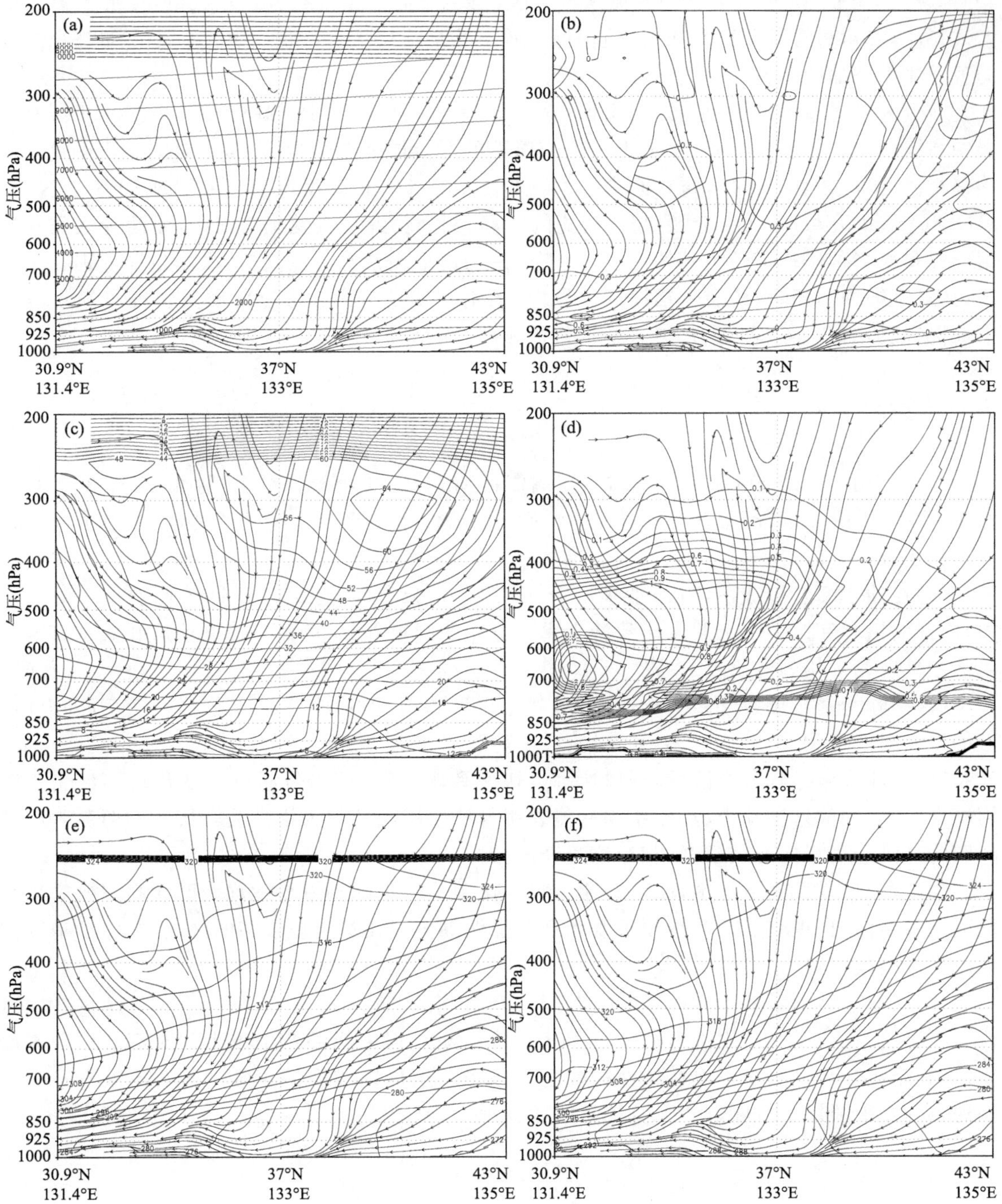

图 11　类似图 8,时间为 2010 年 3 月 21 日 17 时

与之前的孤立沙尘区基本位于相同的高度和纬度，且具有相似的形态特征，从沙尘传输的连续性分析，该沙尘应该来自之前的平流层中的孤立沙尘区。其不位于平流层中，可能是由于沙尘在偏北气流作用下逐渐脱离平流层所致。

急流、比湿与位涡揭示的对流顶下落的情况与之前类似，不再赘述。需要说明的是，40°N，8～10 km(250～350 hPa)的孤立沙尘区，与之对应的比湿＜0.1 g/kg，更接近平流层的大气特征，显示尽管其脱离平流层，但仍保留着一些平流层大气特征。这进一步表明其与之前的孤立沙尘区的同源性。

沙尘区风速垂直切变也与之前类似，对于31°—37°N，4～9 km高度(600～300 hPa)对流层中高层沙尘区，及31°—40°N，1.5～2 km以下的对流层低层沙尘区，与之对应的风垂直切变均很小。从位温、相当位温分布来看，31°—37°N，4～9 km高度(600～300 hPa)沙尘区，及31°—40°N，1.5 km以下沙尘区，与之对应的位温、相当位温垂直变化均很小，而两者之间同样为位温和相当位温梯度较大的锋区，与上节讨论的情形类似。可见，在本次沙尘暴过程各个阶段，在对流层中高层、对流层低层的沙尘层中，风垂直切变、层结、湿度等要素特征均显示其与中性混合层相伴，而这种动力、热力结构特征显然具有普遍性。

5 结论

综合利用CALIPSO星载激光雷达探测资料和MM5数值模拟结果，对2010年3月19—22日影响我国北方的大范围强沙尘暴过程不同阶段沙尘垂直分布特征及其动力、热力结构进行了初步分析，结果发现：

(1)在沙尘暴发展成熟阶段，沙尘层分布于2～9 km(850～250 hPa)的几乎整个对流层中，冷锋前抬升和锋后下沉并未导致沙尘垂直分布的明显差异，反而是锋面前后旺盛垂直混合使沙尘呈现相对均匀的垂直分布。

(2)在沙尘暴扩展及远距离传输阶段，沙尘层明显分为两层，大体上分别位于700 hPa以下的对流层低层和600～300 hPa的对流层中高层。

(3)在沙尘暴过程各个阶段，弱风速垂直切变和弱位温、相当位温垂直变化均与沙尘层密切配合，沙尘层的这种动力、热力结构特点表明与沙尘层伴随的为中性混合层。而两沙尘层之间则为高风速垂直切变及位温、相当位温锋区。

(4)在沙尘暴发展过程中，高空急流、位涡、比湿等要素的垂直变化均表明在高空槽后部与急流下落相伴，出现明显的对流顶折叠和高空位涡下传。在对流顶较高的区域，主体沙尘区向上的扩展也较高，反之则较低。

(5)在本次沙尘暴各个阶段，在40°N，7～9 km沿纬向一线，均出现一小范围孤立沙尘区位于平流层中(或平流层附近)，表明我国北方沙尘暴过程中能够产生沙尘的对流层—平流层输送，并能够在平流层中形成持续性的沙尘传输带。这可以成为沙尘气溶胶对流层—平流层输送及其在平流层中传输的一个直接的监测证据。

参考文献

[1] 王伏村,许东蓓,王宝鉴,等,河西走廊一次特强沙尘暴的热力动力特征分析. 气象,2012,**38**(8):950-959.

[2] 王丽娟,赵琳娜,寿绍文,等,2009 年 4 月北方一次强沙尘暴过程的特征分析和数值模拟. 气象,2011,**37**(3):309-317.

[3] 郭本军,刘莉,黄丹萍,等,激光雷达对一次沙尘天气探测与分析. 气象,2008,**34**(5):52-57.

[4] 郭铌,蔡迪花,韩兰英,等,MODIS 沙尘暴判识方法与业务系统. 气象,2009,**35**(1):102-107.

[5] Ravetta F, Ancellet G, Kowol S J, et al. OPzone temperature and wind field measurements in a tropopause fold: comparison with a mesoscale model simulation. *Mon Wea Rev*. 1999, **127**: 2641-2653.

[6] 杨健,吕达仁. 东亚地区一次切断低压引起的平流层、对流层交换数值模拟研究. 大气科学,2003,**27**(6):1031-1044.

[7] Andreas Stohl, Heini Wernli, Paul James et al. A new perspective of Stratosphere-Troposphere Exchange. DOI: 2003, 10.1175/BAMS-84-11-1565.

[8] Li Chengcai, Shu Hailong, Qinghong Zhang. Evidence of East Asia dust aerosols transported to the stratosphere. In prepared, Private communication.

[9] 姜学恭、陈受钧、顾润源,等. 沙尘暴过程中沙尘对流层—平流层输送的数值模拟初步分析. 气象学报,2012,**70**(6):1223-1234.

[10] Galani, E., D. Balis, P. Zanis, et al. Observations of stratosphere-to-troposphere transport events over the eastern Mediterranean using a ground-based lidar system. *J. Geophys. Res.* 2012, **108**(D12), 8527, doi: 2003, 10.1029/20 02JD002596.

[11] Chen S J, Kuo Y H et al. The effect of dust radiative heating on low-level frontogenesis, 1995, *J. Atmos. Sci.* **52**: 1414-1420.

[12] Carlson T N, Benjamin S G. Radiative heating rates for Saharan dust. *J. Atmos. Sci.* **37**: 1980, 193-213.

[13] 姜学恭,沈建国. 内蒙古两类持续型沙尘暴的天气特征. 气候与环境研究,2006,**11**(6):702-712.

[14] Anthes R A and T. T. Warner. Development of hydrodynamics models for air pollution and mesometeorological studies. *Mon. Wea. Rev*, 1978, **106**: 1045-1078.

Analysis on the Characteristics of Vertical Structure of Sand and Dust in a dust storm process based on the CALIPSO data

JIANG Xuegong[1] CHEN Shoujun[2] YUN Jingbo[1]

1. Inner Mongolia Autonomous Regional Meteorological Observatory, Hohhot 010051;

2. Department of Atmospheric Sciences, School of Physics, Peking University, Beijing 100871

Abstract By using CALIPSO(Cloud-Aerosol Lidar and Infrared Pathfinder Satellite Observations) data and MM5 numerical simulation output data, the characteristics of dusty layers vertical structure and its accompanying dynamic and thermal structure in a severe dust storm occurred from 19 to 22, March, 2010 in North China were studied. The result shown: In the mature period of dust storm, the dusty layer distributed almost in the whole troposphere from 2 to 9 km. The vertical mixture caused by the cold frontal uplifting and the sinking after the front induced a uniform distribution of dust particles. After that, in the long distance transportation period of dust, the dusty layer separated to two layers clearly, located at the lower troposphere (below 700 hPa) and the middle upper troposphere (from 600 to 300 hPa) respectively. In whole periods of the dust storm process, the weak vertical change of wind speed, potential and equivalent potential temperature closely coordinated with the dusty layers. It indicates that the neutral mixed layer maintained in the dusty layer. Between the two dusty layers, the frontel region of potential and equivalent potential temperature appeared. Meanwhile, the evident tropopause fold and obvious declining of large potential vorticity appeared. According with it, when the tropopause sustained at a higher altitude, the dusty layers were expanded to a higher altitude also. Otherwise, the dusty layers were lower. It should noted that in the dust long distance transportation stage, a banding area located at 40°N from 7 km to 9 km appeared in the stratosphere. It indicates that in this dust storm process, parts of the dust particles were transported from troposphere to stratosphere. And it also formed a continued dust transportation belt in the stratosphere. It can be regarded as a straightforward and observational evidence to prove the troposphere-stratosphere transportation of dust aerosol and the transmission of dust aerosol in stratosphere.

Key words: dust storm, dust particles, vertical structure, CALIPSO data

沙尘暴过程中沙尘对流层—平流层输送的数值模拟初步分析*

姜学恭[1] 陈受钧[2] 顾润源[3] 陈 磊[4]

1. 内蒙古自治区气象台,呼和浩特 010051;2. 北京大学物理学院大气科学系,北京 100871;
3. 山东省青岛市气象局,青岛 266003;4. 南京信息工程大学,南京 210044

摘 要:在初步明确东亚沙尘气溶胶对流层—平流层输送监测事实的基础上,利用观测资料、NCEP再分析资料以及基于中尺度天气模式MM5的数值模拟方法,对一次蒙古气旋沙尘暴过程中沙尘对流层—平流层输送问题进行了初步分析,结果表明:斜压不稳定是本次蒙古气旋发展的主要强迫要素,伴随气旋发展成熟,高空切断低涡的形成引导高空急流下落并诱发对流顶折叠和高空位涡下传。对流顶折叠区呈漏斗状,底部达500 hPa左右。高空急流产生近似垂直的下落,并在高空切断低涡的南侧和东侧达到最强。在对流顶折叠区周边的300～500 hPa间,上升气流与低涡区偏西、偏南、偏东气流叠加,或水平横穿折叠的对流顶,或斜升并准垂直地穿过下落的对流顶到达平流层,且随时间的推移,空气质点能够进一步抵达平流层深部(100 hPa)。轨迹分析表明,沙尘天气区对流层低层的空气质点,在气旋涡旋上升气流的驱动下呈气旋式盘旋上升,并在对流层高层形成分支,一支穿过对流顶到达平流层,并在平流层向下游进行反气旋式的螺旋运动,另一支则留在对流层高层并向下游进行准水平的气旋式螺旋运动。在高空位涡下传过程中,主要产生平流层到对流层的净输送;高空位涡停止下传之后则出现对流层到平流层的净输送,且强度随时间呈现随指数型增长。这一特征有利于形成更强的沙尘对流层—平流层输送。

关键词:沙尘暴,沙尘气溶胶,对流层—平流层输送,数值模拟

1 引言

沙尘气溶胶是气候和生态环境的重要影响因素。沙尘气溶胶能够吸收和反射太阳光,减少地面接收到的太阳辐射,导致大气冷却,产生抵消温室效应的作用。沙尘暴形成后,大部分沙尘粒子在沙尘源区附近沉降,其余部分被高空气流输送至遥远的下游地区,通过远距离传输,沙尘气溶胶在大气中的滞留时间和影响范围大大增加,其辐射强迫效应也相应增强。因而沙尘远距离传输是沙尘暴产生气候效应的关键环节。

有关研究估计,全球每年约有10～30×10^8 t沙尘气溶胶注入大气,其中8×10^8 t来自亚洲。在亚洲沙尘暴多发季节—春季,沙尘气溶胶70%来自东亚地区[1,2],其中蒙古及我国北方沙漠区为主要沙尘源区。东亚沙尘气溶胶约半数在源区及其附近沉降,其余主要部分被扬升至3～10 km的对流层并沿40°N附近的带状区域向下游输送,分别在东亚内陆、太平洋西岸、太平洋及北美大陆沉降。PACDEX(PACific Dust EXperiment)研究报告指出,沙尘气溶胶经太平洋到北美洲的远距离传输已经成为全球扩展范围最广和最严重的污染事件,使得春季的太平洋成为全球最大的污染影响海区[3]。东亚沙尘远距离传输在全球气候变化中扮演越来越重要的角色。

* 资助项目:国家自然科学基金项目(40965005)。

沙尘传输的高度也是影响沙尘传输及其气候效应的关键环节。通常情况下，沙尘传输层的高度越高，沙尘传输的距离越远，沙尘在大气中的滞留时间也越久。另外，PACDEX 研究发现，在亚洲和太平洋地区，对流层中高层气溶胶常常侵入到北太平洋中纬度对流云系中，气溶胶与云系的相互作用改变了云系的微物理和辐射特征，导致云辐射——这一温室效应的主要制约因素——对温室效应的抵消作用出现很大的不确定性[3]，因而太平洋地区的气溶胶可能成为影响全球气候变化的敏感因素。而沙尘是太平洋地区大气气溶胶的重要成分，特别是在春季，这进一步说明沙尘气溶胶对于气候变化的重要影响。

需要指出，在特定情况下，如果沙尘气溶胶能够穿过对流层顶到达平流层，则沙尘粒子在平流层滞留时间可以长达 4 天到 2 周，并在一个纬度带内完成环绕地球的传输[3,4]。这种情况下，沙尘气溶胶的气候效应会大大增强。邱金桓等[5]基于激光雷达探测资料研究发现，我国北方强沙尘暴后，北京上空 4～10 km 高度范围内出现气溶胶粒子层，峰值位于 8 km 左右。而 SAGE(Stratospheric Aerosol and Gas Experiment)观测发现北太平洋气溶胶传输层的高度可以达到 15 km 左右。这些研究一定程度揭示了沙尘对流层—平流层输送的可能性。

沙尘气溶胶穿过对流层顶到达平流层表明沙尘上扬过程中产生了对流层—平流层输送，这一过程是对流层—平流层交换(STE)的重要部分。在中高纬度地区，由温带气旋和深对流引发的对流层—平流层输送过程具有重要的气象学意义，原因在于这一输送过程的不可逆性[6]。在温带气旋发展过程中，对流层中高层常形成切断低涡，引发高空锋区加强和对流顶折叠，因而有利于产生对流层—平流层输送[7,8]。输送过程由强烈发展的地面气旋暖区一侧快速上升气流驱动，能够将行星边界层中的空气输送至平流层低层并在平流层中形成传输[6~8]。

在东亚地区，蒙古气旋是沙尘暴的主要天气影响系统之一。作为东亚地区影响最大、发生最频繁的温带气旋，在蒙古气旋的旺盛发展阶段，常常形成对流层顶的折叠，相当多的研究表明其对流层顶下落的高度可以达到 500 hPa 以下。一些有关 STE 的研究也表明蒙古气旋过程中能够产生明确的对流层—平流层输送[9]。可见，东亚地区具备了沙尘气溶胶产生对流层—平流层输送的客观条件，而国内外目前在该领域的研究还是空白。为此，本文拟针对一次典型的蒙古气旋沙尘暴过程，综合利用天气观测资料、中尺度数值模拟输出结果，对蒙古气旋沙尘暴过程中沙尘气溶胶的对流层—平流层输送问题进行初步的探讨。

2 沙尘气溶胶对流层—平流层输送的若干观测事实

明确东亚沙尘气溶胶对流层—平流层输送的观测事实是开展本文研究工作的基础。常规观测手段通常难以捕捉到 UTLS 区域(上对流层—下平流层)的大气气溶胶状况，因此，卫星、激光雷达等成为这类事实的重要监测手段。Li Chengcai 等利用 2000—2005 年间 SAGEⅡ气溶胶观测资料及东亚地区的 98 次沙尘天气过程资料，对比分析了 268 个沙尘暴日与多年平均的常态大气的气溶胶消光系数差异[10]。结果表明，在平流层低层，0.525 μm 和 1.020 μm 波段沙尘暴日大气消光系数明显大于常态大气(图 1)，且 0.525 μm 波段更为显著(通过信度大于 90%的 T 检验)。说明在东亚沙尘暴天气观测中，沙尘气溶胶常常能够完成对流层—平流层输送并到达平流层低层。

另外，陈勇航等利用 NASA 的 CALIPSO 卫星(气溶胶激光雷达和红外开拓者卫星观测)资料，研究了 2007 年 3 月 28 日到 4 月 2 日我国北方一次强沙尘暴过程沙尘气溶胶的垂直分布特征[11]，结果表明，在 34.2°N，93.6°E—46.05°N，97.39°E 轨迹范围内(我国内蒙古西部、蒙古西南部)大量的云与气溶胶并存(图 2a)，气溶胶高度大约在海拔 8～13 km。从其中一点的

(94.46°E,37.19°N)的退偏比可以看到(图2b),在8 km以上,退偏比达到40%~100%之间,说明其中颗粒的不规则程度相当大,在这样的天气条件下,这些气溶胶无疑是微小的沙尘粒子。

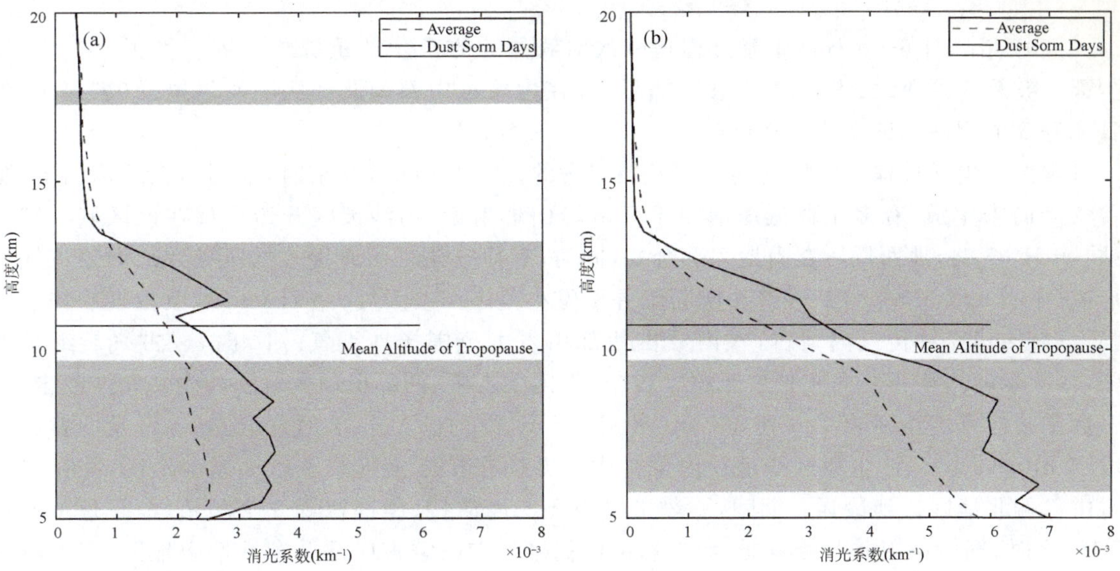

图1 东亚区域(30°—50°N,70°—130°E)2000—2005年98次沙尘天气过程(实线)和6年平均(虚线)的气溶胶消光系数(0.525 μm(a)、1.020 μm(b))对比[10]

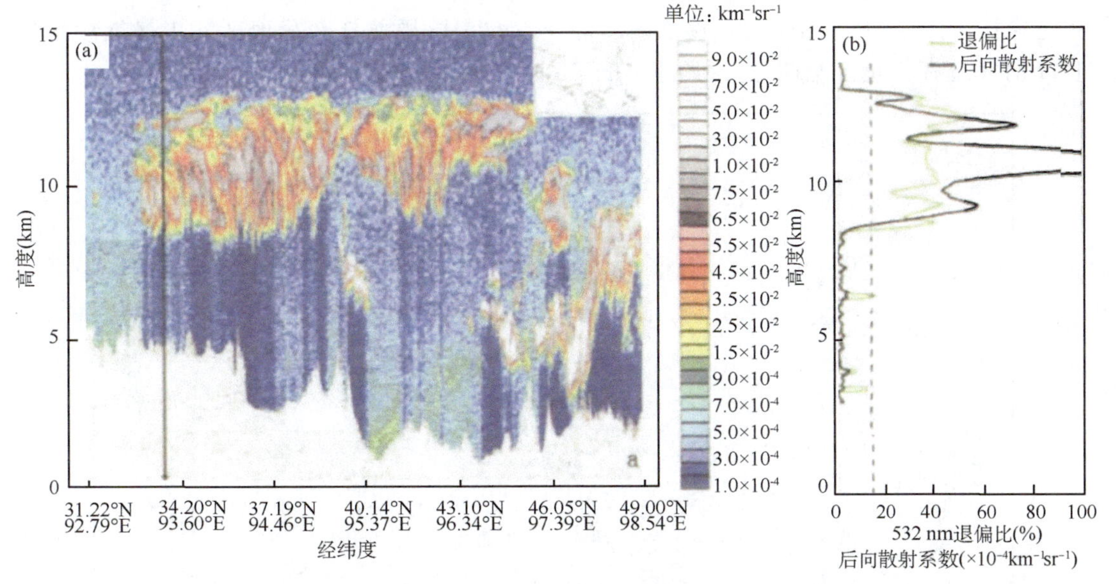

图2 2007年3月29日3:40 532 nm激光总后向散射系数垂直剖面及37.17°N,94.46°E退偏比与总后向散射系数单点廓线。其中,a总后向散射系数垂直剖面,b退偏比和总后向散射系数单点廓线[11]

可见,不论是典型沙尘暴天气个例,还是沙尘暴过程的多年平均状况,都表明在东亚沙尘暴过程中,沙尘气溶胶能够通过对流层—平流层输送到达平流层低层。这些事实为本文以下的分析研究奠定了基础。

3　天气过程概述

2001 年 4 月 6—8 日沙尘暴过程是一次对我国北方产生严重影响的灾害性天气过程，沙尘天气波及我国西北、华北、东北多个省区，并在内蒙古中部偏北地区引发黑风暴（满都拉、朱日和等 5 个测站能见度达到 0 m）。

本次沙尘暴过程是蒙古气旋强烈发展引发的。5 日 20 时（北京时，下同），在地面图上，蒙古气旋尚未生成，有多个弱低压散布于萨彦岭山地附近，扬沙天气开始出现在该区域个别测站。6 日 05 时，地面低压在山地东部合并，同时，有冷空气自西北侵入低压导致地面蒙古气旋形成（中心 995 hPa），气旋冷锋影响新疆并引发小范围扬沙天气。6 日 08 时，蒙古气旋快速增强，气旋冷锋东移侵入内蒙古，在内蒙古西部引发大范围扬沙天气。14 时，蒙古气旋继续增强，沙尘暴随之增强并向东扩展至内蒙古中部，内蒙古中、西部出现大范围沙尘暴和强沙尘暴，多数测站能见度达到 200 m 到 300 m（图 3）。6 日 20 时—7 日 02 时，气旋进入爆发性发展阶段（6 hPa/(6 h)），沙尘暴区稳定在内蒙古中部并向东扩展至内蒙古东部偏南地区，并在内蒙古中部偏北地区达到最强。期间，黑风暴出现在二连浩特、满都拉等 5 个测站（能见度 0 m）。7 日 08 时，气旋锢囚并开始减弱，沙尘暴主体仍维持在内蒙古中部及东部偏南地区，强度略有减弱，同时，浮尘、扬沙、沙尘暴天气开始出现在内蒙古东北部地区及黑龙江、吉林等地。7 日 11 时—20 时，蒙古气旋维持缓慢东移并减弱，沙尘暴区仍主要出现在内蒙古中东部偏南地区，同时，内蒙古东北部地区、吉林、黑龙江等地沙尘暴逐渐增强、范围扩大。8 日 02—05 时，随蒙古气旋减弱，内蒙古地区沙尘天气逐渐减弱停止，而吉林、黑龙江部分地区沙尘暴继续维持至 8 日 17 时才逐渐停止。至此，这次始于 6 日 05 时的强沙尘暴过程结束。

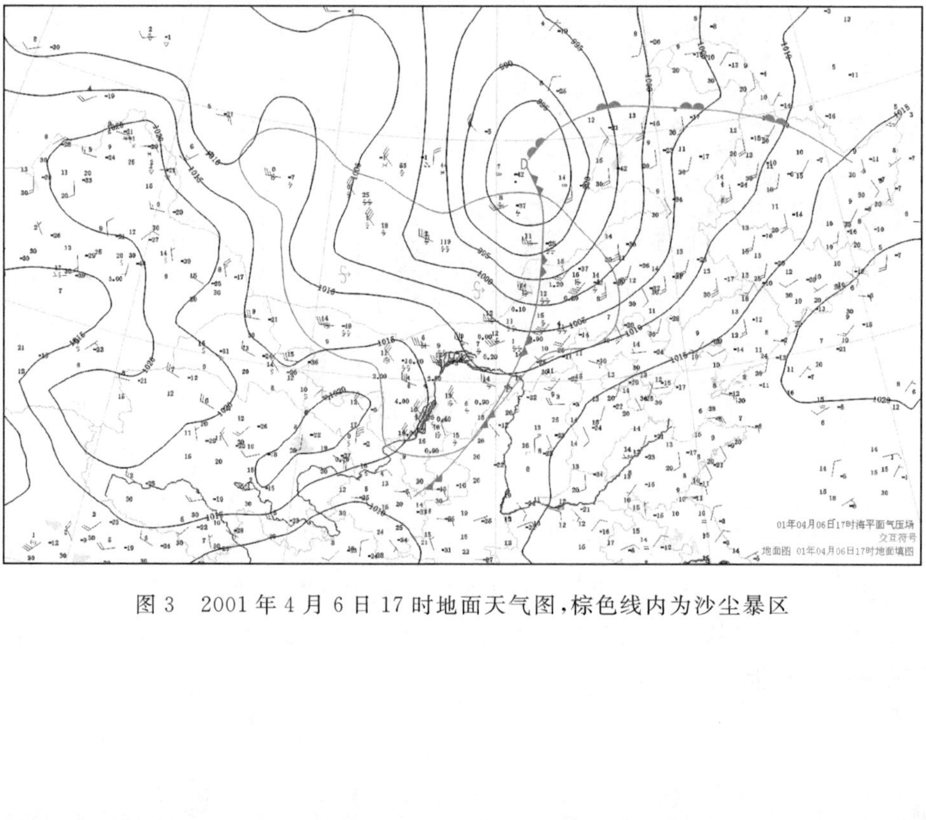

图 3　2001 年 4 月 6 日 17 时地面天气图，棕色线内为沙尘暴区

4　模式和方案

本文对对流层—平流层输送的模拟采用中尺度数值预报模式 MM5。MM5 较多地应用于对中尺度天气过程的预报和模拟，国内外一些研究者也利用 MM5 对对流层顶附近的天气过程进行过成功的模拟研究，如 Ravetta 等[12]利用 MM5 成功的模拟了一次对流层顶折叠事件，模拟的对流顶附近的风场、温度场与观测结果较为吻合。杨健等[9]同样利用 MM5 对 2000 年春季一次切断低压引发的对流层—平流层交换过程进行了成功的模拟研究。可见，MM5 具备了模拟研究对流层—平流层输送过程的条件。

模式采用两重网格嵌套，粗网格格点数 110×90，中心位于(114°E，53°N)，格距 45 km，细网格格点数 193×112，格距 15 km，模式层顶 10 hPa，垂直方向分为 32 层。物理过程采用 Blackadar 高分辨率行星边界层方案和云辐射方案。模式背景场为每日 4 次的 2.5°×2.5° NCEP 再分析资料，模拟时段为 2001 年 4 月 6 日 08 时—10 日 08 时，模拟时间 96 小时。

5　实况和模拟的对比

图 4、5 为 NCEP 再分析和 MM5 模拟结果的对比。首先，从图 4a 500 hPa 高度对比可见，MM5 模拟的低涡位置、强度与 NCEP 再分析结果较为一致，闭合中心为 528 dagpm。两者差异在于 NCEP 再分析的低涡上游高脊略强于模拟结果。从图 4b 500 hPa 温度对比可见，低涡区冷中心范围、强度总体上较为接近，再分析冷中心强度略强于模拟结果。从图 5 500 hPa、300 hPa 风场对比来看，在低涡区及其南侧西风急流区，两者的风速差异均小于 4 m/s，与高度、温度的情况类似，在低涡区上游，两者有 4～8 m/s 的差异。综上可见，MM5 对本次过程高空低涡发展、演变过程进行了较好的模拟。以下将利用 MM5 的模拟结果对本文关心的对流层—平流层输送问题进行进一步的分析和讨论。

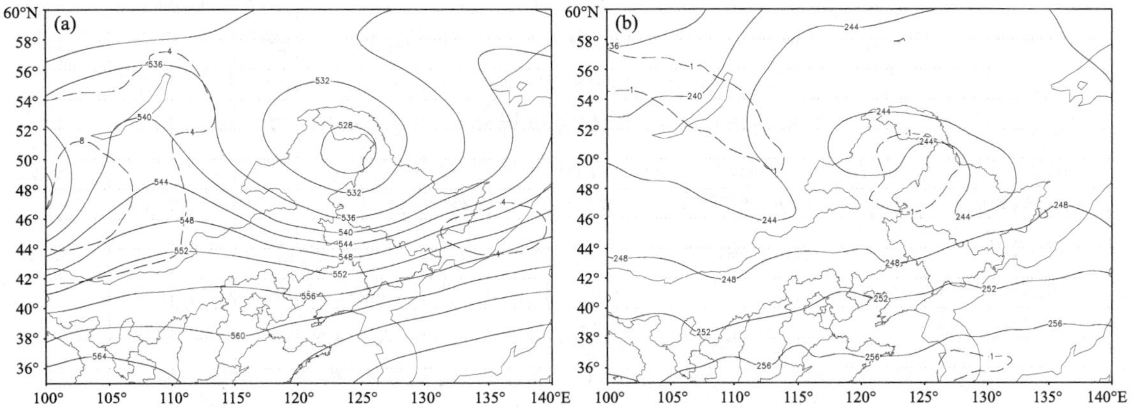

图 4　NCEP 再分析 500 hPa 高度(a)、温度(b)(实线)及其与 MM5 模拟结果差值(虚线)。其中，a. 时间为 2001 年 4 月 8 日 08 时；b. 时间为 2001 年 4 月 7 日 20 时(单位 dagpm，K)

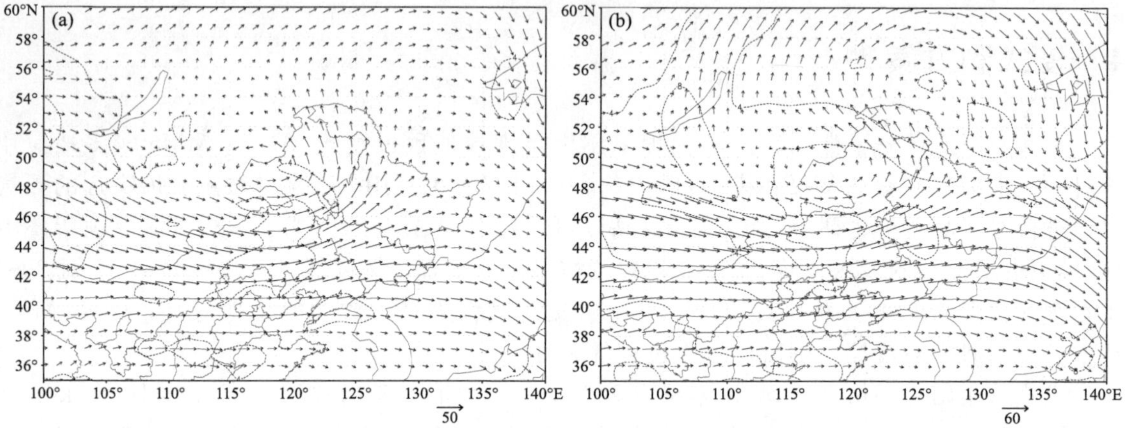

图 5　2001 年 4 月 7 日 08 时 NCEP 再分析 500 hPa(a)、300 hPa(b)风场及风速与 MM5 模拟结果差值(虚线)(单位 m/s)

6　蒙古气旋发展和演变过程

本次过程沙尘暴的形成和发展与蒙古气旋的发展演变过程密切相关(图 6)。5 日 20 时，高空槽位于阿尔泰山以西，冷空气在此堆积，并在 500 hPa 形成－36℃的闭合冷中心，明显落后于高度槽。6 日 08 时，冷空气前沿越过山地，并在地面形成蒙古气旋，中心位于 48°N，104°E，强度 991 hPa(图 6a)。6 日 20 时，温度槽与高度槽接近反位相，呈现典型的斜压结构，有强冷空气平流到高度槽中；高空 600 hPa 以下形成切断低涡，地面气旋已加强至 984 hPa。6 日 20 时—7 日 02 时，气旋移动变缓，强度快速增强(7 日 02 时为 978 hPa)。7 日 08 时，继续有弱冷平流注入低槽底部，对流层中高层相继形成切断低涡(500 hPa 至 300 hPa)。此时，地面气旋锢囚，气旋强度达到 977 hPa(过程最低，图 6b)。7 日 20 时，系统移速加快，低涡区斜压性减小，地面气旋强度明显减弱(图 6c)。8 日 08 时，对流层大部低涡区冷中心消失，低涡强度减弱但仍继续维持，地面气旋继续减弱东移(图 6d)。8 日 17 时，地面沙尘天气全部结束，但对流层大部低涡和地面低压仍维持，直至 9 日 08 时，500 hPa 以下低值系统才完全消失。

综上可见，对流层中低层的斜压强迫是本次过程蒙古气旋的发展的主要影响因素，相关的研究也揭示了这一特征[13]。阿尔泰—萨彦岭山地对低层冷空气的阻滞则使得蒙古气旋的发展经历从缓慢发展到剧烈爆发的过程[14]。前文分析清楚地表明，斜压不稳定首先在对流层低层发展，并在 700 hPa 以下形成切断低涡，之后，切断低涡自下而上扩展，到 7 日 14 时 250 hPa 形成切断低涡。在系统减弱消亡过程中，首先，沙尘天气随地面气旋的消失而结束，而后高空低涡自下而上减弱消失，至 9 日 08 时，500 hPa 以下低涡消失(400 hPa 低涡继续维持)。可见，在系统发展和减弱的整个过程中，这种自下而上扩展的特征表明低层斜压强迫的主导作用。另外，对流层中高层的切断低涡维持时间明显久于对流层低层，这与高空位涡下传对气旋发展的影响有关[13]。这在下文中将进一步阐述。

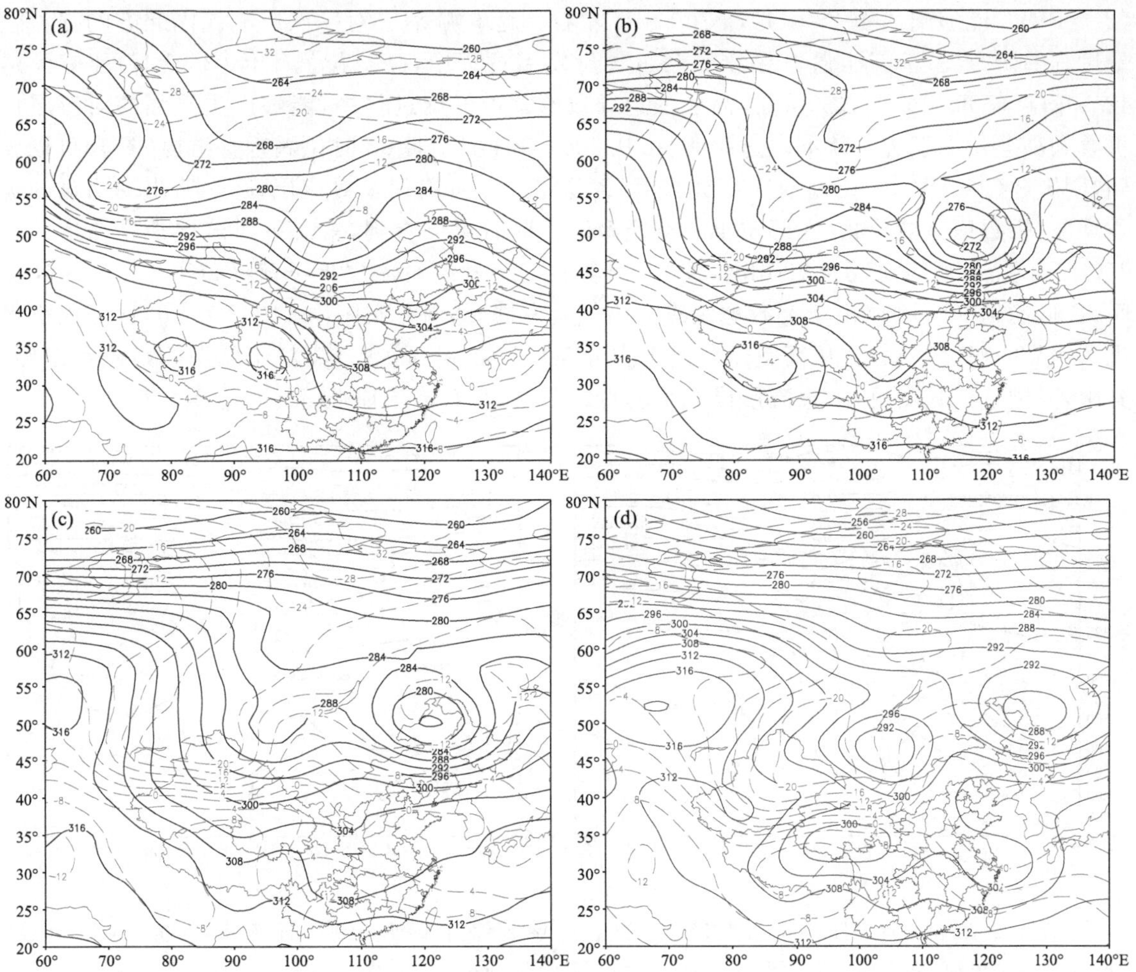

图 6　2001 年 4 月 6 日 08 时(a)、7 日 08 时(b)、7 日 20 时(c)、8 日 20 时(d)NCEP 再分析 700 hPa 高度(实线)、温度(虚线)

7　对流层顶折叠和高空位涡下传

位涡是表征气柱旋转程度的物理量,可以简单理解为单位厚度气柱的绝对涡度。极地平流层是地球大气的位涡源,表现为从极地平流层随着纬度的减小和高度的降低,大气位涡迅速减小。在天气系统发展过程中,往往伴随着高空槽(涡)的加深,而在这一过程中,高空低槽、高空急流、对流层顶的演变是协调一致的,高空低槽的位置影响高空急流的高度和位置,而高空急流的高度又决定着对流层顶的高度,这样,高空槽的加深意味着将使该区域的对流层大气产生向下、向低纬度的压缩,从而能够引导极地平流层的高位涡向下及向低纬度扩展。而位涡无疑会对大气产生旋转强迫,因此,伴随对流层顶折叠的高空位涡下传是中高纬度天气系统发展的一个重要的强迫因素。

在蒙古气旋发展过程中,高空位涡下传常常作为重要的动力强迫因素之一。杨健等[9]研究表明,由高空切断低涡触发的蒙古气旋发展过程中,伴随对流层顶折叠的高空位涡下传

(2.5 PVU)能够到达40°N附近和700 hPa左右。赵琳娜等[15]对一次影响我国北方的强沙尘暴过程的研究发现,斜压强迫导致的蒙古气旋发展过程中,高空位涡下传(2 PVU)能够到达42°N附近,500 hPa左右。可见,高层强迫导致的蒙古气旋过程中,能够产生更强的对流层顶折叠和高空位涡下传。就本次过程而言,从2001年4月7日08时位涡垂直结构(图7)可以看到,阴影区(2<PV<3 pvu,有关研究表明,PV≥2 pvu可以代表对流层顶[4])所出现明显的折叠现象,折叠区呈漏斗状,底部位于高空低涡区。同时可以看到,常态下对流层顶高纬地区在250 hPa左右,低纬地区为200 hPa左右,而随着气旋发展对流层顶下落区位于40°—53°N之间,最低达到500 hPa(由于上游地区有另一气旋发展,纬向分布图中110°—125°E区域内也出现了明显的对流层顶)。从高空全风速垂直分布可以看到,高空急流位于低涡的南侧和东侧,其中,南侧急流强度更强,急流形成几乎垂直的下落,与对流层顶下落区接近重合。

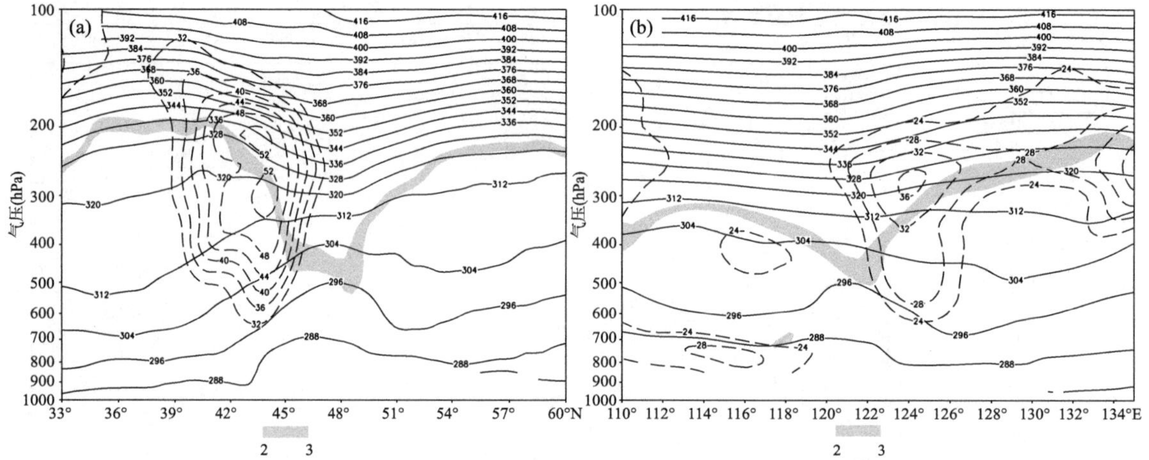

图7 2001年4月7日08时沿121°E经向(a)和48°N纬向(b)垂直剖面图,其中,实线为位温,单位K,虚线为全风速,单位m/s,阴影区为位涡(2 pvu<PV<3 pcu)

以上分析可以得出:首先,本次蒙古气旋发展过程中,出现明显的对流层顶折叠现象。如果将高空低涡发展引发的急流下落视为对流层顶折叠的驱动因素,那么驱使对流层顶下落的动力主要集中在低涡的南侧和东侧。其次,陈受钧等研究表明,从发生机理角度,蒙古气旋可以分为地形主导和地形影响两种类型:前者以大气对地形导致的锋面变形的响应为主要触发因素;后者以对流层中高层形成切断低涡为重要特征,高空位涡下传是气旋发展的触发因素[16,17]。后者无疑能产生更强的对流层顶折叠和高空位涡下传,即前文文献所讨论的天气过程[18]。而本次过程正如前文分析的情况,天气系统自下而上的发展显示低层强迫占有主导地位,因此与赵琳娜等[15]研究的同属于第一种类型。对流层顶折叠的程度也较相似。下文的分析表明,尽管如此,对流层顶折叠仍成为本次过程实现沙尘对流层—平流层输送的适宜条件。

8 对流层—平流层输送的结构和强度特征

图8、9为2001年4月7日08时沿经向和纬向穿过低涡区的垂直剖面图。从图8a可以看到,在对流层顶折叠区即低涡区南侧的800～300 hPa间(对流层低层流场呈现明显的湍流特征),南风与上升气流叠加形成向北爬升的斜升气流,在到达300～500 hPa附近与折叠的对

流层顶南支相遇，并产生穿过对流层顶的运动，其中，在 400～500 hPa 间，对流层顶接近准水平形态，该气流以由南向北斜升的姿态穿过对流层顶到达平流层；而在 300～400 hPa 之间，该气流由斜升气流转为准水平气流，自南向北穿过折叠的对流层顶进入平流层；进入平流层后，该气流仍为自南向北的斜升形态，其中，部分气流重新穿过折叠的对流层顶北支返回对流层中，其余则维持在平流层中的斜升运动。从图 8b 纬向剖面图可以看到，偏西气流与上升气流叠加，同样在大约 300～500 hPa 之间穿过折叠的对流层顶，到达平流层。进入平流层后，气流维持向东斜升的形态，部分气流穿过对流层顶重新回到对流层中，其余部分则停留在平流层中。

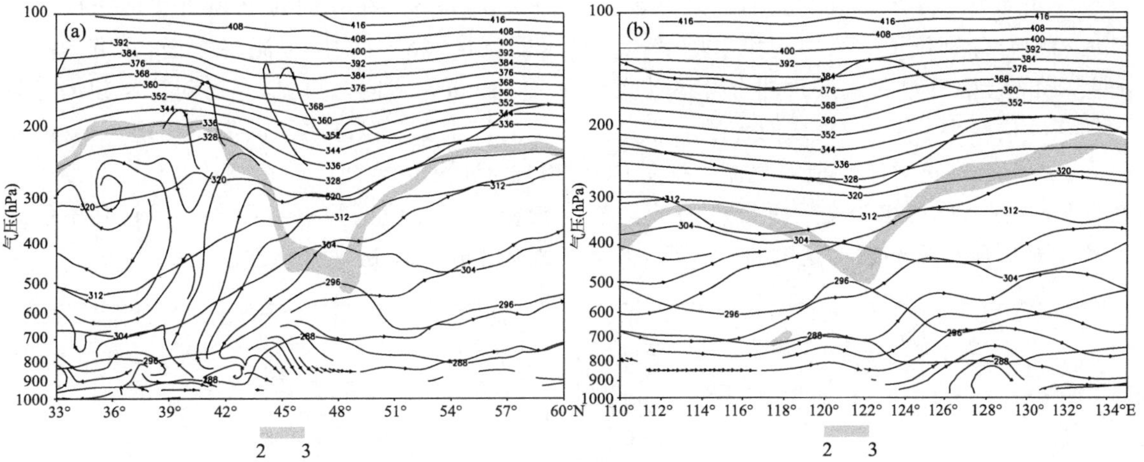

图 8 同图 7，其中，流线分别为 v, w 合成(a)和 u, w 合成(b)，垂直速度 W 单位为 cm/s

图 9 为 7 日 14 时(图 8 的 6 小时后)沿 121°E 和 50°N 的经向和纬向剖面图，可以看到，首先，在经向剖面中并未出现水平方向穿越对流层顶的气流，在漏斗状对流层顶折叠区底部，经向水平气流很弱，上升气流以准垂直方向穿过对流层顶到达平流层中，在平流层中，上升气流继续爬升，一直达到 100～200 hPa 之间的平流层深部。从纬向剖面图可以看到，由于 50°N 附近为低涡中心附近偏北一侧，因此，在漏斗状对流层顶折叠区的底部，上升气流与弱的偏东气

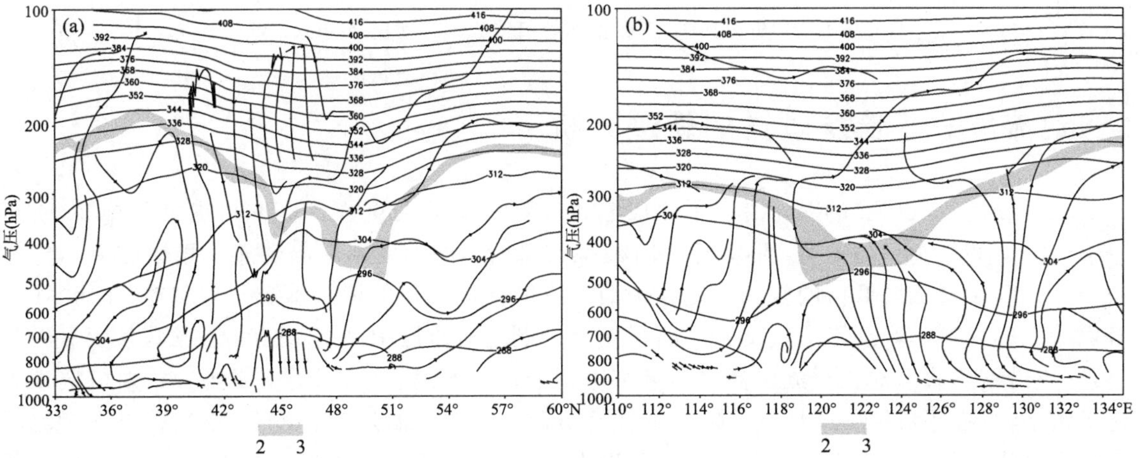

图 9 2001 年 4 月 7 日 14 时沿 121°E 经向(a)和 50°N 纬向(b)垂直剖面图，其余同图 7

流叠加穿过准水平的对流层顶到达平流层中，且该气流同样在平流层中继续爬升并到达平流层深部(100～200 hPa 间)。而春季中纬度 100 hPa 高度大约在 15 km 左右，这与 SAGE 观测到的气溶胶达到的最大高度是一致的。另外，由于在该蒙古气旋的上游地区有另一蒙古气旋生成，沿 50°N 经向 90°—130°E 的宽阔区域内形成了不同程度的对流层顶下落(250 hPa 以下)，在对流顶下落区域同样出现了气流穿过对流层顶进入平流层的现象。

为考察穿过对流层顶的上升气流强度，图 10 给出了 7 日 08 时垂直速度的强度分布，可以看到，在经向对流层顶折叠区南侧 400 hPa 附近有 12 cm/s 的上升运动中心，北侧同样在 400 hPa附近为 6 cm/s 的上升运动中心，这两支上升气流均未直接穿过对流顶。而对流层顶折叠区底部直接穿过对流层顶的上升气流强度约为 6 cm/s。在纬向对流顶折叠区东西侧上升运动中心分别位于 500 hPa 和 700 hPa，同样未能直接完成对流层平流层输送，在宽阔的对流层顶下落区，穿过对流层顶的上升运动大约在 2～3 cm/s 之间。可见，穿过对流层顶的上升运动强度均较弱。因此，完成对流层—平流层输送的气流主要以准水平的姿态出现。另外，图 8 显示准水平穿过对流层顶的气流主要集中在 300～400 hPa 之间。而春季 350 hPa 高度大约在 8 km 左右，这与相关文献所得出的沙尘气溶胶粒子层峰值区是一致的[5]。尽管其不能作为沙尘对流层—平流层输送的直接证据，但无疑强化了这一输送的可能性。

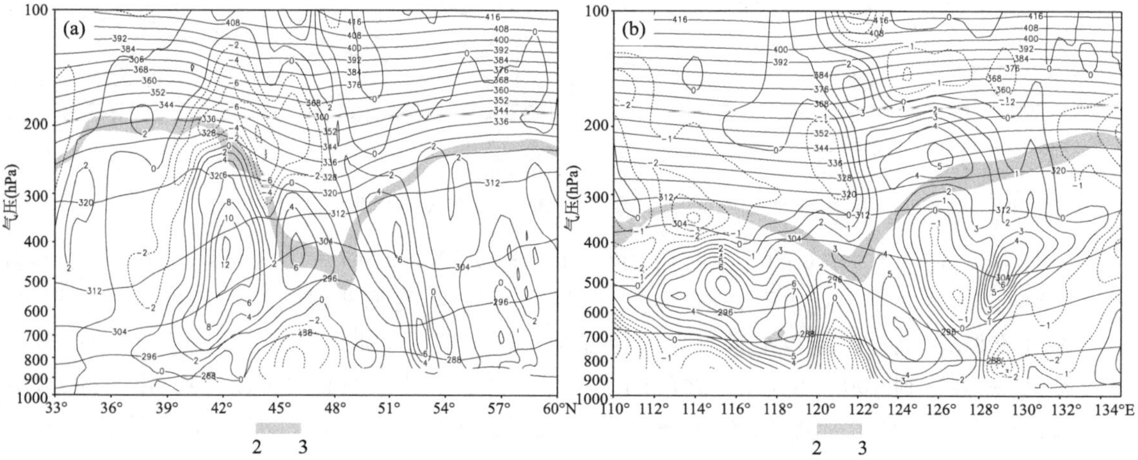

图 10　同图 7，其中，正负相间等值线为垂直速度，单位 cm/s

9　对流层—平流层输送的轨迹分析

为了分析空气质点的对流层—平流层输送过程中的运动轨迹，图 11 给出了利用 2001 年 4 月 6 日 08 时—10 日 08 时 MM5 逐小时模拟结果所做的前向和后向轨迹。可以看到，首先，位于蒙古高原对流层低层的空气质点在偏西气流的作用下伴随地形蜿蜒向东移动，而后，在气旋涡旋上升气流的驱动下沿气旋式弯曲盘旋上升到达对流层中高层，之后，形成两条明显的分支，一支沿气旋式弯曲在对流层高层向下游进行准水平的涡旋运动；另一支则穿过对流层顶到达平流层，并在平流层中以反气旋的方式在平流层低层向下游运动。可见，空气质点在对流层高层和平流层低层表现出完全不同的运动形态特征，即对流层高层为气旋式，而在平流层为反气旋式。这与有关研究的结果是一致的。从图 4 可以看出，在 50°—54°N，125°—130°E 区域内,由于天气系统随高度后倾,500 hPa 为低涡控制的气旋式气流区,而 200 hPa 平流层低层为高脊控制的反气旋式气流区,这一定程度决定了空气质点的运动轨迹。其深入的动力学原因还有待进一步研究。

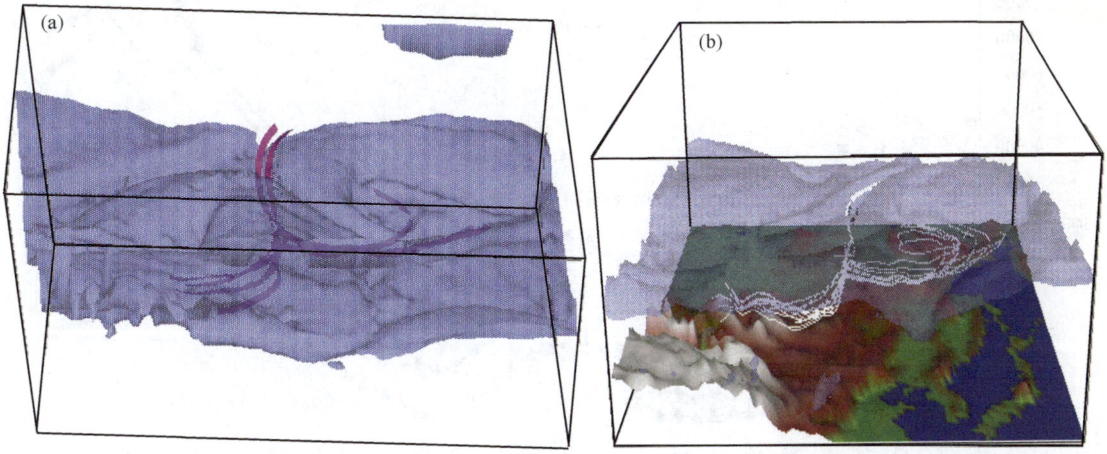

图 11　空气质点轨迹,资料为 MM5 模拟的 2001 年 4 月 6 日 08 时—10 日 08 时逐小时输出结果。图中,曲面为等 PV 面(PV=2 pvu),其中,亮色和暗色分别标明轨迹位于等 PV 面上方和下方(a. 前向;b. 后向)

10　沙尘对流层—平流层输送的估算

以下对沙尘气溶胶对流层—平流层的输送量作了初步的估算。由于没有沙尘气溶胶浓度分布的资料,仅对发生在对流层顶附近的对流层—平流层交换的体积进行估算。输送体积的计算利用以下公式:

$$\text{Area} = (\text{cdiff}(lon, x)/2 \times \pi/180 \times \cos(\text{lat} \times \pi/180) \times 6.37e^3) \times (\text{cdiff}(lat, y)/2\pi/180 \times 6.37e^3) \tag{1}$$

其中,Area:网格面积,单位 km^2;cdiff:中央差分;π 取 3.1416;$6.37e^3$:为地球半径。

计算方法为:取随时间移动的低涡区约 10 个经纬度矩形区域,以 PV=2 pvu 作为对流层顶,当某时刻对流层顶附近 $w>0$ 时(上升运动)记为对流层—平流层输送,并将垂直速度乘以面积微元得到输送体积微元,然后针对矩形低涡区域进行积分,得到输送体积;反之则为平流层—对流层输送。

图 12 给出了利用上述方法得到的低涡从发展到成熟期间对流层—平流层交换的体积变化,可以看到,6 日 23 时—8 日 03 时之间为平流层—对流层净输送,之后则为对流层—平流层净输送,且输送体积呈现指数型增长。而地面气旋在 7 日 02 时达到最强,6 日 23 时前后气旋接近成熟,而高空位涡下传主要出现在气旋成熟之后[13]。可见,平流层向对流层的输送过程是与对流顶折叠和高空位涡下传过程一致的。另一方面,此时沙尘粒子正处于从对流层低层到对流层高层的长传过程中,因而此时平流层—对流层的净输送并未影响沙尘的对流层—平流层输送。之后,在气旋减弱且高空位涡下传停止之后,对流层—平流层输送开始显著增强,而此时沙尘粒子也已经完成从对流层低层到高层的上传,因而能够产生明显的沙尘对流层平

流层输送。这一现象也表明气旋发展过程客观上有利于沙尘对流层—平流层输送的发生。

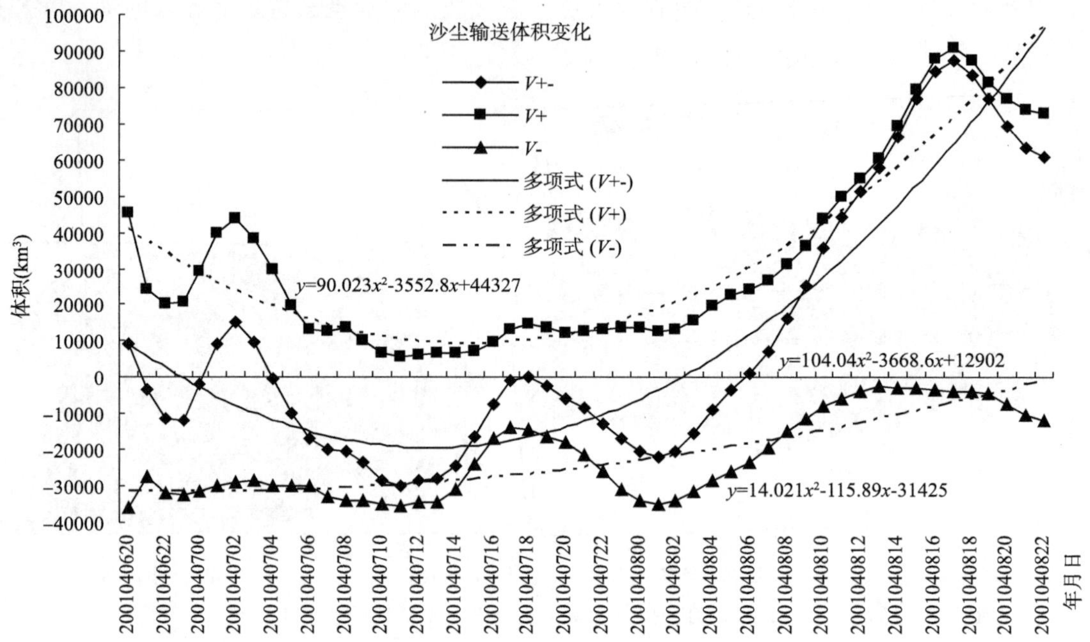

图 12　对流层—平流层交换体积随时间演变，其中 V＋代表 $w>0$ 的体积变化，V－代表 $w<0$，V＋－为两者之和

11　小结

在初步明确东亚沙尘气溶胶对流层—平流层输送监测事实的基础上，利用天气观测资料、NCEP 再分析资料并结合 MM5 数值模拟结果及轨迹分析方法，对一次蒙古气旋引发的强沙尘暴过程中沙尘的对流层—平流层输送问题进行了初步分析，结果表明：

(1)本次过程中蒙古气旋的发展是由对流层中低层的斜压不稳定首先触发的，在气旋的成熟阶段，高空切断低涡的形成及高空位涡的下传使气旋的发展进一步增强，但在气旋发展过程中并不具有主导作用。这类过程不是最有利于产生对流层—平流层输送的天气过程。

(2)伴随着气旋的发展成熟，对流层切断低涡呈现自下而上的扩展，并引导高空急流的下落和对流层顶的折叠。对流层顶折叠区呈明显的漏斗状，底部到达 500 hPa 左右，期间伴随高空急流接近垂直的下落，急流核位于高空低涡的南侧和东侧，且南侧最强。

(3)蒙古气旋发展过程中的上升运动中心位于 500 hPa 和 700 hPa，且并未直接穿过折叠的对流层顶。在对流层顶折叠区周围的 300～500 hPa 间，上升气流与低涡区偏西、偏南、偏东气流叠加，或以水平气流的形态横穿折叠的对流层顶，或以斜升的形态准垂直地穿过折叠的对流层顶底部到达平流层。随着时间的推移，完成对流层平流层输送的空气质点能够进一步抵达平流层深部(100 hPa)。

(4)在对流层顶折叠区的外围，宽广的对流层顶下落区同样出现穿过对流层顶的上升气流，但其强度较对流层顶折叠区明显偏弱。

(5)轨迹分析表明，沙尘天气区对流层低层的空气质点，在向下游运动的过程中，在气旋涡旋上升气流的驱动下呈气旋式盘旋上升，并在对流层高层形成分支，一支穿过对流层顶到达平流层，并在平流层中向下游进行反气旋式的螺旋运动，另一支则留在对流层高层，并向下游进行准水平的气旋式螺旋运动。

(6)在高空位涡下传过程中，主要产生平流层到对流层的净输送；高空位涡停止下传之后则出现对流层到平流层的净输送，且强度随时间呈现随指数型增长。这一特征有利于形成更强的沙尘对流层—平流层输送。

参考文献

[1] Penner J E, Andreae M, Annegarn H, et al. Aerosols, their direct and indirect effects, in IPCC 2001: *Climate Change, The Scientific Basis, Contribution of Working Group I to the Third Assessment Report of the Intergovernmental Panel on Climate Change* [Houghton, J. T., et al., (eds.)], Cambridge University Press, Cambridge, United Kingdom and New York, NY, USA, 2001.

[2] Zhao T L, Gong S L, Zhang X Y, et al. A simulated climatology of Asian dust aerosol and its trans-Pacific transport. Part I: Mean climate and validation. *Jour. Clima.* 2006, **19**(1): 88-103.

[3] Stith J, Ramanathan V, et al. PACific Dust EXperiment(PACDEX), 2006, Third Release.

[4] Andreas Stohl, Heini Wernli, Paul James et al. A new perspective of Stratosphere-Troposphere Exchange. DOI: 2003, **10**. 1175/BAMS-84-11-1565.

[5]邱金桓，郑斯平，董其荣，等. 北京地区对流层中上部云和气溶胶的激光雷达探测. 大气科学，2003，**27**(1): 1-7.

[6] Holton J R, Haynes P H, et al. Stratosphere-troposphere exchange. *Rev. Geophys.*, 1995, **33**: 403-439.

[7] Wirth V and Egger J. Diagnosing extratropical synoptic-scale stratosphere-troposphere exchange: A case study. *Quart. J. Roy. Met. Soc.*, 1999, **125**: 635-655.

[8] Stohl A, Wernli H, James P, Bourqui M, Forster C, Liniger M A, Seibert P, and Sprenger M. A new perspective of stratosphere-troposphere exchange. *Bull. Am. Meteoro. Soc.*, 2003, 1565-1373.

[9] 杨健，吕达仁. 东亚地区一次切断低压引起的平流层、对流层交换数值模拟研究. 大气科学，2003，**27**(6): 1031-1044.

[10] Li Chengcai, Shu Hailong, Zhang Qinghong. Evidence of East Asia dust aerosols transported to the stratosphere. 2012, In prepared, Private communication.

[11] 陈勇航，毛晓琴，黄建平，等. 一次强沙尘输送过程中气溶胶垂直分布特征研究. 中国环境科学，2009，**29**(5): 449-454.

[12] Ravetta, F, G Ancellet, S J Kowol, et al. OPzone temperature and wind field measurements in a tropopause fold: comparison with a mesoscale model simulation. *Mon Wea Rev.* 1999, **127**: 2641-2653.

[13] 姜学恭，沈建国，刘景涛，等. 地形影响蒙古气旋发展的观测和模拟研究. 应用气象学报，2004，**15**(5): 601-611.

[14] 姜学恭，沈建国. 导致一例强沙尘暴的若干天气因素的观测和模拟研究. 气象学报，2003，**61**(5): 606-620.

[15] 赵琳娜，赵思雄. 一次引发华北和北京沙尘暴天气的快速发展气旋的诊断研究. 大气科学，2004，**28**(5): 723-735.

[16] Chen S J, Egger J and Han W. Altai-Sayan lee cyclogenesis: A case study. *Beitr. Phys. Atmos.*, 1992, **65**: 13-22.

[17] Han W, Chen S J, and Egger J. Altai-Sayan lee cyclogenesis: numerical simulations. *Meteor. Atoms. Phys*, 1995, **55**: 125-134.

[18] 杨健，吕达仁. 平流层—对流层交换研究进展. 地球科学进展. 2003，**18**(3): 380-385.

A Preliminary Numerical Analysis on the Troposphere-Stratosphere Transportation of Dust Aerosol in a Dust Storm Event

JIANG Xuegong[1] CHEN Shoujun[2] GU Runyuan[3] CHEN Lei[4]

1. Inner Mongolia Autonomous Regional Meteorological Observatory, Hohhot 010051;
2. Department of Atmospheric Sciences, School of Physics, Peking University, Beijing 100871;
3. Qingdao Meteorological Bureau, Shandong province, 266003;
4. Nanjing University of Information Science & technology, Nanjing 210044

Abstract Based on a confirmation of troposphere-stratosphere transportation(TST)of dust aerosol in East Asia, the TST of dust aerosol in a Mongolia cyclongenesis induced super dust storm process was conducted by using observationaland NCEP reanalysis data and through numerical simulation method. The result shown: The Baroclinic instability is the dominant forcing causing the Mongolia cyclongenesis. Following the maturation of cyclone, the formation of upper level cut low led the subsiding of upper level jetstream, which induced the tropopause fold and the downward transportation of upper level potential vorticity. The folded tropopause was showing funnel-shaped with its bottom approaching near 500 hPa. The upper level jet stream conducted quasi-vertical subsiding with its core locating the south and east sides of the cut low. Around the tropopause fold region and between 300 hPa and 500 hPa, the resultant wind between the ascending motion and the western, southern and eastern wind, passed through the folded tropopause into stratosphere horizontally or aslant upwards(through the bottom of folded tropopause). As time goes by it approached the deep part of stratosphere(100 hPa). The trajectory analysis shown that the air particles at the lower troposphere in the areas occurring dust weather were raised upward driven by the spiral ascending motion of cyclonic area. The dust aerosol transportation belt split two branches when it got to the upper lever of troposphere. One passed through the folded tropopause into stratosphere and then conducted anticyclonic movement downwards within it. The other retained in the troposphere and maintained cyclonic movement downwards. A net Stratosphere-Troposphere Transportation occurred accompanying with the process of download transportation of upper level potential vorticity. After that, a net Troposphere-Stratosphere Transportation took place and its strength exhibit an exponential growth. This characteristic is favorable for the Troposphere-Stratosphere Transportation of dust aerosol.

Key words: dust storm, dust aerosol, troposphere-stratosphere transportation, numerical simulation